Early Railway Chemistry and its Legacy

Early Railway Chemistry and its Legacy

Colin A. Russell and John A. Hudson
The Open University

RSCPublishing

ISBN: 978-1-84973-326-7

A catalogue record for this book is available from the British Library

© Colin A. Russell and John A. Hudson 2012

All rights reserved

Apart from fair dealing for the purposes of research for non-commercial purposes or for private study, criticism or review, as permitted under the Copyright, Designs and Patents Act 1988 and the Copyright and Related Rights Regulations 2003, this publication may not be reproduced, stored or transmitted, in any form or by any means, without the prior permission in writing of The Royal Society of Chemistry or the copyright owner, or in the case of reproduction in accordance with the terms of licences issued by the Copyright Licensing Agency in the UK, or in accordance with the terms of the licences issued by the appropriate Reproduction Rights Organization outside the UK. Enquiries concerning reproduction outside the terms stated here should be sent to The Royal Society of Chemistry at the address printed on this page.

The RSC is not responsible for individual opinions expressed in this work.

Published by The Royal Society of Chemistry,
Thomas Graham House, Science Park, Milton Road,
Cambridge CB4 0WF, UK

Registered Charity Number 207890

For further information see our web site at www.rsc.org

Printed and bound in the United States of America

Preface

To work in a chemistry laboratory is always a privilege, but to occupy one overlooking the West Coast Main Line might be regarded as a prize, a coincidence or just a nuisance, depending on one's view of railways. This was the experience of one of us [CAR] who, while trying to do chemistry, found it an occasional diversion to see mighty legends like *The Royal Scot* go thundering past a mere few yards away from the window (especially as these were the very last days of steam). Less glamorous but much slower and noisier were trains hauled by elderly goods engines, conveying to Scotland soda ash and salt from Cheshire, all vital to the chemical industry. It never occurred to him that there might be an essential connection between the work of a chemical laboratory and the shiny ribbon of steel snaking away into the distance carrying these trains to far-off destinations. He could not have been more wrong.

Railway historians and enthusiasts will be surprised to learn that chemists played an important part in the development of the railway industry in Britain. Chemists themselves are well aware of the many and wide-ranging applications of their discipline, but the fact that their predecessors were involved in the technological development of railways will come as a surprise to many. The present book is the first detailed study of this important interaction, and as such it is designed to appeal not only to chemists and to railway enthusiasts, but also to engineers and historians of science and

---

Early Railway Chemistry and its Legacy
By Colin A. Russell and John A. Hudson
© Colin A. Russell and John A. Hudson 2012
Published by the Royal Society of Chemistry, www.rsc.org

technology. It is part of a wider picture embracing science and technology in general and the railways.[1]

Writing for such a wide readership has its pitfalls. For this reason, where unfamiliar ideas are introduced in either the field of railways or chemistry, a simple explanation is offered to those for whom they might be unfamiliar. However, we have tried to avoid too many technical terms which, as chemists, we might otherwise take for granted. *Early Railway Chemistry and its Legacy* is concerned with the role of chemistry in the British railway industry from its beginnings in the early 19th century up to the Grouping of the railways of 1923 into the Great Western Railway, Southern Railway, London and North Eastern Railway and London, Midland and Scottish Railway. Later developments are dealt with more briefly.

Perhaps a little should be said about our historical approach to this subject, mentioning two issues in particular. First, we know that many historians believe that chemistry, like all sciences, is affected and determined by the social environment in which it develops; this seems to us undeniable by the facts. A small minority of these writers (rarely including chemists) additionally hold that chemistry is *entirely* a social construct. We are not of their number, and have tried to give an accurate factual account of what actually happened, leaving others to speculate as they wish. Our second point is slightly related, and concerns the social *effects* of science. We agree that chemistry has had the most enormous social consequences in a whole range of fields, and here in the history of railways its contribution has been massive (which is why the book exists in the first place). Chemistry's function here has turned out to lie in the development of the railway system in the UK, and thus to reinforce the social effects that railways have had: population mobility, seaside holidays, accelerated business deals, and many others. In other words, here is a classic example of how a science has related to a technology, in a multitude of ways that no one imagined in the first days of railways, in Britain of the 1830s, and this technology has been the means of social change.

The book began some years ago with the discovery of papers that had so far been hidden from view. It became the subject of a PhD research project by one of us [JAH][2], working in conjunction with the other [CAR]. Although two authors are involved, the present book is very much a joint venture. The authors have frequently met, and each has read, and commented on, the draft chapters of the other. The present book is the first full length study of this important topic, although two papers on aspects of railway chemistry have recently been published by one of us.[3]

A wide variety of source material has been used for the first time, and we want to express our great indebtedness to many people, including especially the late Mr Mike Hall for bringing fresh material to our attention and for encouraging us in this project. Very sincere thanks are due to Dr Alastair Gilchrist, formerly technical director of British Rail Research, who assisted one of us [JAH] in cataloguing an archive of railway technical papers at Derby. We are grateful to the former owners of that archive, Scientifics Ltd, for providing access and allowing a number of items to be photocopied. This important collection of documents is now in the care of the National Railway Museum at York. Local and national archives have provided much material, hitherto unknown. We also acknowledge the important papers of Sir Edward Frankland, owned by Mr Raven Frankland and now deposited in the John Rylands Library of the University of Manchester. Many other individuals have helped us, and we are glad to acknowledge the work of the RSC Publications Division, and above all the patience and assistance of our wives, Judy Hudson and Shirley Russell.

Every effort has been made to obtain the necessary permissions for with reference to copyright material. Should there be any omissions in this respect we apologise and shall be pleased to make the appropriate acknowledgements in any future edition.

REFERENCES

1. C. A. Russell, *Proc. Inst. Mech. Eng.*, Part F, 1998, **212**, 201.
2. J. A. Hudson, *Chemistry and the British Railway Industry 1830–1923*, unpublished PhD thesis, The Open University, 2005.
3. J. A. Hudson *Manchester Memoirs*, 2010, **147**, 42; *idem, Int. J. History Eng. Technol.*, 2011, **81**, 274.

Contents

Abbreviations	**xiii**
Chapter 1	
Railways without Chemists?	**1**
References	9
Chapter 2	
***Rocket* and its Hidden Chemistry**	**10**
2.1 Iron	15
2.2 Non-ferrous Metals	17
2.3 Lubrication	18
2.4 Water	18
2.5 Fuel	20
References	22
Chapter 3	
Chemistry: A New Force in Society and in Railway Development	**24**
3.1 The First Chemical Revolution: Coal, Iron and the Rest	24
3.2 The Second Chemical Revolution: Atoms and Molecules	29
3.3 A Scientific Social Revolution: Enter the Chemist!	37
References	41

Early Railway Chemistry and its Legacy
By Colin A. Russell and John A. Hudson
© Colin A. Russell and John A. Hudson 2012
Published by the Royal Society of Chemistry, www.rsc.org

Chapter 4
Testing the Waters: The Early Railways and their First Chemical Consultants   **43**

4.1	The Dawn of the Railway Age	43
4.2	The Problem of Boiler Scale	44
4.3	Early Water Analyses for Railway Companies	46
4.4	Water Analysis Becomes Standard Practice	50
4.5	Water Treatment	54
4.6	Conclusion	56
References		56

Chapter 5
More Work for the Chemical Consultants   **59**

5.1	Introduction	59
5.2	Fuel Analysis	59
5.3	The Preservation of Timber	61
5.4	Air Quality in Tunnels	65
5.5	The Swansong of the Chemical Consultant	68
5.6	Conclusion	69
References		69

Chapter 6
A New Breed of Railwayman: The Railway Chemist   **73**

6.1	The Testing of Steel	73
6.2	Chemists at Crewe	75
6.3	Chemists after Crewe	79
6.4	The Railway Chemists: Their Public Image	88
References		90

Chapter 7
The Railway Chemists as Materials Testers   **94**

7.1	Introduction – Testing		94
7.2	Specifications and Standards		96
7.3	Analytical Methods in General		97
7.4	Metals		99
	7.4.1	Steel	99
	7.4.2	Copper	101
	7.4.3	Other Metals, and Materials used in Foundry Work	102

7.5	Fuels	103
7.6	Oils for Lighting	105
7.7	Water	105
	7.7.1 Water for Locomotives	105
	7.7.2 Water for Domestic Consumption	108
7.8	Lubricants	109
7.9	Paints	114
7.10	Miscellaneous Materials	116
7.11	Analysis of Goods Tendered for Carriage, and Claims for Damage	117
7.12	Other Activities	121
7.13	Conclusion	122
References		123

Chapter 8
Research on Railway Issues — **128**

8.1	Introduction	128
8.2	The Background to Railway Research	128
8.3	Metals	129
8.4	Lubricants and Lubrication	138
8.5	Water Treatment	143
8.6	Minor Research Projects	145
8.7	Conclusion	148
References		148

Chapter 9
The Railway Chemists in Collaboration — **151**

9.1	Collaboration or Competition?	151
9.2	The Railway Clearing House and the Carriage of Dangerous Goods	152
9.3	The Railway Chemists' Committee	154
9.4	New Work for the Committee	157
9.5	The Railway Chemists in World War I	161
9.6	Conclusion	162
References		162

Chapter 10
The Enduring Legacy of the Railway Chemists **164**

10.1	Grouping and After	164
10.2	Work Accomplished	167
	10.2.1 Water	167
	10.2.2 Lubricants	169
	10.2.3 Adhesion	172
	10.2.4 Materials Examination and Development	173
	10.2.5 Environmental Work	175
10.3	Conclusion: The End of the Story?	177
References		178

Subject Index **181**

Abbreviations

CCA	Cheshire and Chester Archives, Chester
DCRO	Durham County Record Office
MCL	Manchester Central Library
NAS	National Archives of Scotland, Edinburgh
NCL	Newcastle Central Library
NRM/DA	An archive of documents formerly held by Scientifics Ltd., Derby, and now housed at the National Railway Museum, York
RFA, OU mf	Raven Frankland Archive, Open University microfilm. Archive now at John Rylands Library, Manchester.
RI MS	Royal Institution, Manuscript Collection
SML	Science Museum Library
TNA	The National Archives, Kew, London
WSRO	Wiltshire and Swindon Record Office, Trowbridge

CHAPTER 1
Railways without Chemists?

The nineteenth century has often been called "The Age of the Railway". It saw the birth of locomotive-hauled railway trains, and the coming of the public railways was one of the most powerful instruments of massive social change that had yet appeared. The subsequent story is mainly confined to Britain, for that is where much of the action took place. The Victorian railway has indeed been the subject of innumerable books, far too many to mention, but by way of example is one splendidly summarised by its title: *Railways and the Victorian Imagination*.[1] Book after book has appeared glamorising the achievements of "the iron road". Indeed, railways touched on most of life at some time. They gave the population a new mobility and a fresh availability of goods from faraway places. They moved coal from its mines to millions of homes, they were largely responsible for the establishment of the annual seaside holiday, and they made possible the mass movement of livestock, even including whole circuses. Railways, alongside heavy manufacture and textile production, are a superb example of the dawning of a new technology which transformed society, and, being individual examples of private enterprise, spawned a huge propaganda industry. Railways are portrayed as (amongst other things) a product of consummate engineering skills, but almost nowhere is there even a mention

Early Railway Chemistry and its Legacy
By Colin A. Russell and John A. Hudson
© Colin A. Russell and John A. Hudson 2012
Published by the Royal Society of Chemistry, www.rsc.org

of indebtedness to any of the sciences, least of all chemistry. Its conspicuous absence from railway propaganda has materially assisted in the belief that it was irrelevant if not totally absent.

At the same time as this happened, Victorian Britain saw what might justifiably called a second "scientific revolution". It was not as far-reaching as the so-called "Scientific Revolution" of the 17th century, but it was important nonetheless. Immense advances took place in some of the older sciences that had long-term effects on chemistry and on technology itself. Alongside electricity, chemistry itself exploded with new ideas, and it would have been surprising if the railway industry had been unaffected. An even bigger surprise lies in the wholesale ignorance as to whether this is the case or not. It is a contention of this book that the unexpected *did* happen, that railways were affected by chemistry, and that without chemistry the railways would not have existed as we know them, or at least they would have been unimaginably different.

There were, of course, some obvious ways in which the opposite trend was true, and chemistry has been heavily beholden to the railway industry. It had often been alleged that the science owed more to railways than *vice versa*. Thermodynamics was indebted to some pioneering studies of the steam engine (though it must be said that most of these early steam engines were of the stationary type, and in many cases had nothing to do with rail traction as such). All this is commonplace. There were other interactions, however, as may be seen exemplified by several well-documented rail journeys by chemists in the 19th century.

At 9.20 am (or as near as possible) on an October morning of 1845, a train of yellow coaches steamed slowly southwards out of Lancaster, the only through train that day to London (Euston). The Lancaster & Preston Railway was conveying among its passengers, a young man, Edward Frankland, who had just completed his apprenticeship in a local chemist's shop (see Figure 1.1). He had had his head stuffed full of new ideas about chemical theory which he had picked up at unofficial evening classes in Lancaster run by two philanthropic doctors. This was the science for him, and one of his benefactors had managed to obtain the promise of work in the Department of Woods and Forests, based in Westminster and headed by the aspiring young chemist, Lyon Playfair, born a mere seven years before Frankland himself.

Figure 1.1 Edward Frankland (1825–1899).

The next five years saw many changes for Frankland, but included the foundation of organometallic chemistry and the first glimpses of a theory of valency, one of the most basic features of chemical thought, both then and now. They also witnessed his acquisition of a Marburg PhD, and the start of a meteoric career that was destined to revolutionise chemistry itself. One remarkable thing was that it all began with that railway journey from Lancaster.[2] Indeed, there was no other practical means of transport that did not take several days, and in later years it was the newly formed network of railways in Britain that made possible the holding of provincial conferences by such organizations as the Chemical Society and the British Association for the Advancement of Science. All science thus profited by railway travel but, if a science is in a state of massive flux as was chemistry, meetings between kindred minds become imperative. In that sense, as we shall see again, chemistry in Britain was peculiarly indebted to the railways. As for Frankland himself, while his vitally important contributions to chemistry are well known, knowledge of his important work for the railway industry has only recently come to light (see Chapters 4 and 5). Yet there is further evidence from overseas as we shall now see.

By 1860, chemistry was in a condition that can only be described as "dreadful". To no one was this more apparent than August Kekulé, professor of chemistry at Ghent and famous for his later hexagon formula for benzene, but whose textbook on organic chemistry gave no less than 19 formulae for one compound, acetic acid. Partly this was a reflection of confusion regarding atomic weights. Accordingly, Kekulé conceived an international conference at the eastern town of Karlsruhe in order to resolve the matter. Unfortunately it was largely unsuccessful, though an Italian delegate, Stanislao Cannizzaro, did propose a solution though it is perhaps too technical to describe here. As delegates left after a few days wrangling, Cannizzaro presented them with a copy of his own course of chemistry which contained this solution. One of the delegates, Lothar Meyer, pocketed a copy "to read on the way home". He was returning to Breslau in the far east of the country (now in Poland and known as Wroclaw). Today travellers fly easily between the two cities, and although we do not know Meyer's exact method of travel, it almost certainly included a long journey by rail. He had time to read and re-read the paper, and he gradually perceived its value. As he said "the scales seemed to fall from my eyes. Doubts disappeared and a feeling of quiet certainty took their place".[3] The discovery was enshrined in his textbook, *Die Modernen Theorien der Chemie*, an influential work written two years later. Whether this enlightenment actually began in a railway carriage, and was completed at home, we cannot be sure. The conference itself, one of the first international conferences of any science, could never have been possible, however, without the burgeoning railway network all over Europe.[4] As a recent scholar has observed, "the steamboat, railway and the telegraph did much to facilitate the formation of national and international communities in science, including those in chemistry".[5] For that reason alone – and there are many others – the debt of chemistry to the railways is immense.

In terms of transit, the growing chemical industry was often reliant on the railways for transport of raw materials (see Figure 1.2) as well as people, though much material was also conveyed by sea or canal. Chemical plants were sometimes served by trains hauled by "fireless" locomotives, well filled with steam but having no fire to replenish them or to ignite flammable material nearby. Service of chemistry by railways continued. In the USA, railways were for many years almost the only long-distance carriers

Railways without Chemists?

Figure 1.2 Allhusen Works, Gateshead, 1907, showing the railway lines serving the chemical company. With permission from Gateshead Council, Libraries and Arts, Gateshead Central Library.

of iron, salt and gunpowder, and they continue to play an important role today. This is true of many industrialised countries in the latter part of the 19th century.

This book focuses on "traffic" in the other direction, *i.e.* chemistry in the service of the railways. For most people, the two are unrelated. Railways today have a huge enthusiast following. Thousands of people, mainly men and boys, maintain intense interest in railway practice and performance, and the steam engine, though banned for regular service by British Rail in 1968, still exerts an almost fanatical fascination. Not all of these aficionados wear anoraks! Many of them have a detailed knowledge that never ceases to astonish. They display a pre-nationalisation enthusiasm that once supported individual railways, fanned into even greater fervour by pre-war books like the *Boys of all Ages* series by W. G. Chapman for the Great Western Railway. This series of books contained an incredible amount of technical detail on all aspects of railway operation and sold over 100 000 copies.[6] The first in the series came out in August, 1927. Nor is there any shortage of relevant literature today, with many periodicals, of which the old-established *Railway Magazine* tops the list. As we write this book, a major manufacturing venture has recently completed work on a new A1 Pacific (4-6-2) steam engine, *Tornado,* to a design by A. H. Peppercorn from just after the war (see Figure 1.3). The manufacture was brilliantly successful. It is intended that it will be used for main line steam haulage of "special" trains, like its Peppercorn predecessors in regular service.[7] Indeed, the first year of service of the steam engine gives every indication that this goal will be fulfilled.

Chemistry can never claim anything like the place of affection accorded to railways, but it has served them in one spectacular way, in occasionally providing the explosive needed for railway cuttings and tunnels. Most of this work was manual, or with mechanised diggers, while tunnels used various "shields". Since the beginning, however, these processes have occasionally been helped by gunpowder, a product of chemical investigations from many centuries earlier, containing the chemicals sulfur, charcoal and saltpetre. In 1865, Alfred Nobel invented dynamite, nitroglycerin absorbed on kieselguhr. Dynamite was applied in the boring of the Mont Cénis tunnel, joining France and Italy, in 1869, and led to a

Figure 1.3 *Tornado*, a new locomotive produced to a previous design, on the North Yorkshire Moors Railway in April 2011. Photography by P. Benham.

five-fold reduction in time compared with the use of gunpowder.[8] Unfortunately, the discovery of dynamite was a little too late to be of use in Britain, having occurred just after the greatest period of railway expansion.

Today chemistry has lost much of its popular glamour. Partly this is because it is much more complex and the new practical applications are rather less obvious than before (except in medicine and allied fields). For one thing, it is now a profession, with trained and highly skilled chemists being paid to do it, so experiments by unrecognised "amateurs" are rightly frowned upon. For another, it is a well-established school subject, and few things diminish affection so much as *having* to learn them!

However, it was not always thus, and the popularity of chemical science knew no bounds in the 19th century. Again, in those days, it was mainly men and boys who went to the greatest lengths to study the subject in such limited spare time as they had. It was so popular that when "science" was mentioned in connection with government exams it almost always meant chemistry. There were reasons for this. First of all chemistry was widely seen as a "useful subject", and one that could benefit mankind by improving working conditions in a multitude of ways. It was incidentally the belief in the power of science and technology that helped to make Britain such a powerful manufacturing nation: "the workshop of the world". Chemistry was also seen as an exciting subject, with fires, explosions and odours of all kinds. Not for nothing did Victorian schoolboys call it "stinks"!

More seriously, today the combined efforts of security authorities and compliant teachers have meant that many of the most spectacular experiments are no longer demonstrated in the lab. What is left is much "tamer" and less likely to set the pulses racing (this, by the way, is about *popular* perceptions, not what real chemistry is actually about).

All of this means that to talk of railways and chemistry in the same breath suggests a certain incongruity, and a connection that in most people's minds has never been made. Yet, as we have mentioned, we believe that without application of chemistry the railways of Britain would be almost unrecognisable today. Subsequent chapters will explore why this should be so, not least by demonstrating that water (for steam engines) had to be analysed so as to improve boiler efficiency and reduce maintenance costs.

Chemistry also pronounced on the difference between cast iron, steel and wrought iron, so important for all kinds of engineering, not least the production of rails.

Such a claim is so far-reaching that it is easy to dismiss it without further thought. A vast amount of new evidence has appeared in the last few years, however, which substantiates our assertion. It came about in the following way.

For some years a detailed investigation into the origins of the chemical profession had suggested that chemical analysis was important for many things, including early railways.[9] Then a chance encounter led one of us [CAR] to explore the matter more fully. Most important was the input of the late Mr. Mike Hall. He was then area chemist at a railway laboratory at Crewe, set up over a century earlier by the London and North Western Railway. Mr. Hall had had unrivalled access to documents about his laboratory, stretching back to its foundation. Amongst the treasures in the Hall collection was a number of laboratory notebooks, several photographs, minutes of the London and North Eastern Railway (LNER) Chemists' Council, many official publications, lists of chemists and above all a typescript history of the Crewe laboratory to show that it is not a "fantasy of travellers' brains". Even so, both the London, Midland and Scottish Railway (LMS) and LNER documents have a few inaccuracies, as was discovered when original material in the archives was examined.

Some of these documents went to the Derby laboratory, and are now in the National Railway Museum at York; we have some originals and copies, and the rest have since disappeared. From all these documents, it became clear that for over 100 years railways had employed their own staff ("railway chemists") to conduct and supervise all kinds of chemical analysis relevant to the running of the railways. Since then it has become apparent that, in a consultant capacity, many independent chemists of great ability had been employed by the railways on an *ad hoc* basis, almost back to the beginnings of the industry.

In the following chapter, something of the chemical challenges facing the first railways will be discussed in connection with that iconic locomotive, *The Rocket*. A chapter then follows which shows how chemistry was ready to rise to the occasion. How this interaction between chemistry and the railways developed is the theme of the succeeding chapters of the book. It is an important, though hitherto largely unknown, case of the relationship between science and technology.

REFERENCES

1. M. Freeman, *Railways and the Victorian Imagination*, Yale University Press, New Haven and London, 1999.
2. Further details of this chemist may be found in C. A. Russell, *Edward Frankland: Chemistry, Controversy and Conspiracy in Victorian England*, Cambridge University Press, 1995 (revised edn 2003).
3. L. Meyer, *Ber. Dtsch. Chem. Ges.*, 1887, **20**, 997.
4. Five years year later Germany had 7826 km of railways, and by 1875 that had increased to an enormous 27 970 km. (N. Davies, *Europe, A History*, Oxford University Press, 1996, p. 1297.)
5. H. Kragh, in *The Making of the Chemist*, ed. D. Knight and H. Kragh, Cambridge University Press, 1998, p. 331.
6. W. G. Chapman, *Locos of the Royal Road*, David and Charles, Newton Abbott, 1987, reprint, publishers' preface.
7. M. Allat, *Railway Magazine*, 2008 (April), 14, and many later references.
8. B. Morgan, *Civil Engineering: Railways*, Longman, London, 1971, p. 141.
9. C. A. Russell, N. G. Coley and G. K. Roberts, *Chemists by Profession*, Open University Press, Milton Keynes, 1977.

CHAPTER 2
Rocket and its Hidden Chemistry

If, as we have suggested, no clear connection has been perceived for a long time between railways and chemistry, it would indeed be interesting to know if this connection has ever existed. It is not difficult to show that railways in their infancy generally managed very well without taking much account of chemistry. Indeed, semi-popular treatments of locomotive development are often devoted almost exclusively to engineering (*e.g.* boiler design, wheel arrangements, *etc.*).[1] However, the intention of this chapter is to expose the chemical issues that were there all the time, waiting to be solved, and *had this been possible sooner* the course of railway history might have been very different. As it happened, half a century elapsed before railways employed their own chemists in any numbers. Meanwhile, chemists occasionally served as consultants making small, but important, contributions to railway development.

Many of those involved very early in the railway industry, whether promoters, directors, managers and even some engineers, were unaware of how science in general, and chemistry in particular, could assist them in their endeavours. There were however, shining exceptions. Foremost amongst them were Isambard Kingdom Brunel and Daniel Gooch of the Great Western Railway (Figures 2.1 and 2.2). They obtained analyses of water and of coke from various kinds of coal; they performed experiments on coke ovens; they evaluated cokes

---

Early Railway Chemistry and its Legacy
By Colin A. Russell and John A. Hudson
© Colin A. Russell and John A. Hudson 2012
Published by the Royal Society of Chemistry, www.rsc.org

Figure 2.1 Isambard Kingdom Brunel (1806–1859).

Figure 2.2 Daniel Gooch (1816–1889). With permission from Swindon Collection, Central Library, Swindon.

in service, measuring fuel consumption and water evaporated per mile in different locomotives; they experimented with various timber preservatives, corresponding with Michael Faraday and others; and they were in contact with John Bethel about trials with his newly introduced creosote oil for timber preservation. Gooch's interest in science is borne out by his appreciation of the work of the local Mechanics' Institute, located at that time within the Swindon works.[2] Of course there were instances of railway engineers not consulting chemists when they should have done, and paying dearly for their omission, especially when they found that they were using water of an unacceptable quality in their locomotives. Lessons were quickly learnt, however, and the commissioning of an analysis of a potential source of water became the norm from the late 1840s onwards.

To examine the potential role of chemistry in the early industry, we shall focus, though not exclusively, on that iconic locomotive, the *Rocket*, designed by George Stephenson (see Figure 2.3) and his son Robert. *Rocket* (see Figure 2.4) was not merely the winner of the famous Rainhill Trials of 1829 (see Figure 2.5) for the most competent design, but in 1830 became the first ever main-line locomotive, regularly hauling passengers on the Liverpool and Manchester Railway. Fortunately, much of *Rocket* still survives and can be inspected in the Science Museum in London. After 140 years in the London museum, Rocket was placed for one year, during a gallery reorganisation in 1999, in the National Railway Museum at York. Here, for the first time, a very detailed examination was made of the whole machine, together with relevant archive research. The research was conducted with meticulous care, and a comprehensive report was recently published.[3] One of the many facts emerging from the study was that *Rocket* had undergone

Figure 2.3 A picture by the 19th Century artist John Lucas; the central male figure (holding a lamp) is George Stephenson, the man on the left is his father and on the right is his son Robert. Also viewable is the early locomotive *Puffing Billy*, and the cottage where Robert Stephenson was born. Although the situation is imaginary, the likeness of the three Stephensons is apparently good.

Rocket *and its Hidden Chemistry*

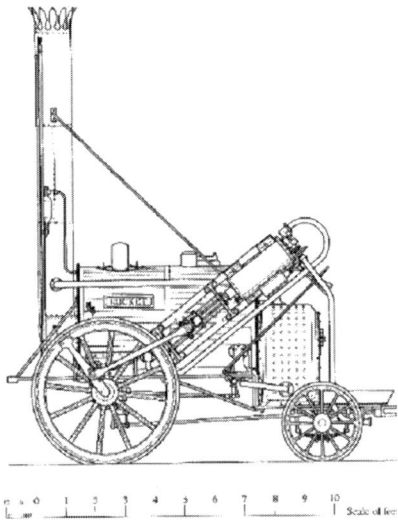

Figure 2.4 *Rocket*, an early diagram. With permission from M. R. Bailey and J. P. Glithero (from their book on the *Rocket*).

Figure 2.5 The Rainhill Trials, October 1829.

several significant modifications since it was first assembled. These were due in part to accidents, changes of function, improvements added over its years in service, and to (undocumented) changes during its stay at the Science Museum. Nevertheless, it was something of an exaggeration when one Victorian engineer announced in 1881 "That the engine has almost been rebuilt is certain".[4] However, we are now in a position, as never before, to understand how the original locomotive was constructed and how it must have looked.

In the late 1820s, chemistry was gaining pace (as the next chapter will show), but it did not seem relevant to the new railways. We have already noted something of the scientific efforts of Brunel and Gooch, but these tended to be the exception. Sociologists might argue that the railway and chemical communities were quite distinct and rarely overlapped. However, too much can be made of this; George Stephenson, for example, was a member of the distinguished Literary and Philosophical Society of Newcastle, an institution to become famous for its chemical members and later a progenitor of the Newcastle Chemical Society.[5] Stephenson was friendly with the alkali manufacturer, James Losh, and other chemists. He became one of the first to use the Literary and Philosophical Society's apparatus, though this was chiefly in connection with his invention of a miners' safety lamp (and a rival to Davy's) rather than in connection with railways or the *Rocket*. The irrelevance of chemistry to railways may have been partly due to separate social units with their own agendas, but was, in large measure, a result of the primitive state of chemical analysis at that time and its sheer difficulty.

As late as 1865, the specifications of the GWR locomotive, *Hackworth*, stated that the cylinders should be made from "a mixture of best Shropshire cold blast and Blaenavon iron", and the boiler barrel was to be made from "Low Moor or other approved plates", while the inside firebox was to be constructed of "copper". The chemical composition of the various metallic materials was hardly considered worthy of a mention. With the chemical analysis of steel, however, this situation began to change.

Shortly afterwards, though well before the 1923 Grouping resulting from the Railways Act of 1921, chemistry began to be applied in an increasing number of areas of railway construction and maintenance. Before we look at specific chemical applications, however, it is important to examine the areas where chemistry is of particular relevance. In a word, it is all about the materials involved in making and running railway engines at around the time of the *Rocket*. It is the chemistry of materials that is the key area of interest, though the term "materials science" did not come about until much later, during the 1950s.[6] What then were the materials where chemical knowledge would have been beneficial?

2.1 IRON

This familiar metal was to become integral to the Industrial Revolution, then beginning to dominate much of Victorian Britain. An abundant element on Earth, iron, as it is today, was extracted from iron ore (by removal of oxygen, usually with charcoal or coke, or other materials rich in carbon). Since the 15th century, the making of iron was facilitated by forcing a blast of air through a hot mixture of iron ore and charcoal, producing a higher temperature and a better reduction of the ore (*i.e.* removal of oxygen). The "blast furnaces" were crude affairs, originally reliant on water power to generate enough pressure on the air. The product from the blast furnaces, fused iron, was run into moulds constructed to look rather like a group of piglets being suckled by their mother. For this reason the product was sometimes called "pig-iron", or otherwise "cast iron". Cast iron was strong, though brittle when struck. We now know that such "iron" contains about 2.2% to 4.5% of carbon, together with other impurities.

Charcoal was eventually replaced by the solid product formed from heating coal in the absence of air, and later from the coal–gas industry. This was another form of impure carbon, and familiar to us as "coke". In fact, it was the shortage of charcoal which was one of the spurs for developing coke smelting and coal-fired puddling. As we shall see in the next chapter, Abraham Darby first smelted iron ore with coke well over 100 years before the Liverpool and Manchester Railway. Although coke smelting was slow to catch on, it was almost universally used by 1830. There was no way that the charcoal-based iron industry could have satisfied the demands of the railways, even in the early days. The replacement of charcoal by coke, and of water power by steam power, had profound social effects. As a result, iron-making, and other industries such as cotton spinning, moved from rural locations to industrial centres, thus causing the rise of cities like Manchester, and hence creating the demand for the railway connecting Manchester and Liverpool.

Another form of iron, wrought iron, was made by roasting molten pig iron in the "finery", so that the carbon in it was less than 0.25%. An improved process, called "puddling", was in use by the 1820s, the product of which could withstand being struck and was malleable, being capable of being rolled and machined into all kinds of useful shapes, *e.g.* piston rods. It could also be hammered to remove some of the remaining slag and give a fibrous structure. Wrought iron was the chief starting material for the *Rocket*, as it had been for its predecessors. When Robert Stephenson began work on railway engines, both forms of iron (wrought and cast) were available to him, and he obviously knew their properties and possibly their approximate chemical composition. He did not select the iron on a chemical basis, however, but rather because of its known mechanical properties. Boiler plates were sometimes obtained from a Staffordshire company, though other items of the same material (*e.g.* for piston rods) were obtained more locally.

For the wheels, cast iron had been employed on other, earlier engines, but they were extremely heavy, and in 1833 the Stephensons replaced these with wooden wheels until wrought iron spokes were successfully developed. Cast iron was used for the driving-axle horns and the rear-support brackets under the boiler. These could be cast into shape with spaces to minimise the weight. The two cylinders were also cast using the same material.

Steel was made in small quantities by the crucible process, and adjusting the carbon content of the wrought iron. At that time this was an expensive procedure, and produced steel of an uncertain quality. Steel was rarely used except in the manufacture of springs. There were four spring-sets for the wheels, to support the boiler. Only later were steel springs incorporated into the buffers. All these iron components had their own particular characteristic chemical compositions, but in the 1830s this was not seen as relevant to the engineers. Nevertheless, as we shall see, it was later to become of supreme importance for the rails on which subsequent trains would run. The rails of early tramways were made of wood. Later, cast iron was used, and by 1830 early railways were employing rails made exclusively from wrought iron. It was the development of the technique of rolling wrought iron rails in 15 foot lengths by John Birkinshaw in 1820 that enabled the development of railways (as opposed to waggonways or tramways). Trevithick's demonstration of a locomotive on the Penydarren tramway in 1804 did not lead to the development of railways at that time due to the cast iron rails which broke under the weight of the locomotive. The locomotive itself (although very primitive by *Rocket* standards) performed well.

Eventually steel became the universal material. It was the composition of steel that engaged the first professional railway chemists, and since then the chemical analysis of this form of iron has been crucially important for the safe running of the railways.

2.2 NON-FERROUS METALS

A quite different metal employed in *Rocket* was copper, a softer and more ductile metal than iron. The copper used was of variable quality, and different samples had different compositions, but this was of less importance than the varying carbon content of commercially available iron. Copper was the material used to construct the firebox of *Rocket* (and other locomotives), having good thermal conductivity and other virtues. In the very earliest engines, the gases from the firebox passed through a pipe in the boiler, thereby heating the water. However the process was very inefficient, and, in the *Rocket*, the Stephensons replaced this single hot pipe with 25 copper tubes providing a much greater heating surface with a vastly improved performance. Robert Stephenson fixed the tubes to the firebox by welding or "clinking", as he put it. The use of a multi-tube boiler was one of the locomotive's great innovations and was continued to be used in Britain up until the end of steam. The bursting or breaking of the copper elements caused considerable problems, however, which might have been avoided by chemical control of the copper. As early as 1838, copper tubes were being replaced by "brass" ones, though what was generally called "brass" (an alloy of copper and zinc) by the middle of the 19th century had more tin than zinc in it, so approximated more closely to our "bronze" (a true alloy of copper and tin). "Brass" was originally used for axle bearings, but by the middle of the 19th century it was superseded by white metal: an alloy of lead, tin and antimony. All these alloys could have very varied compositions, but their chemical analysis was ignored until quite a bit later in the century.

Many years passed before copper once again became a metal of great interest, not least to William Dean, locomotive superintendent of the Great Western Railway. For some years, the engineers had complained that copper no longer stood the wear and tear in tubes and fireboxes. In part, this probably reflected the greater strain and higher temperatures in later locomotives, but probably it was also a question of composition. At that time copper had to be highly refined for electrical use, and it was thought that the absence of small amounts of arsenic and other metals may have been responsible for its mechanical weakness. At last, chemical examination was used to solve the problem.[7]

2.3 LUBRICATION

All machinery generates friction, and this adds to the demands of the prime mover. Hence efforts were always being made to reduce this waste of energy. In the days of the *Rocket*, this was done partly by grease. Some grease was made from tallow and palm oil, simmered with soda solution. A photograph of a bearing running surface from *Rocket* shows the oil feed holes and distribution grooves. These were fed from an oil reservoir. The oil for *Rocket* was probably supplied by W and E Joy.[8] Mineral oils were also tried, especially in America, and by the 1860s these had largely replaced grease; well-designed oil boxes were simpler than the old grease boxes. To complicate matters, many later steam engines had oil which came into contact with steam, or at least hot water; if the oils contained esters (such as palm oil), hydrolysis would occur and the oil would lose its lubricating properties, and in the worst cases, corrosive fatty acids would form. Different parts of the engine needed different types of lubricant. The chemistry of running a locomotive was complex and initially poorly understood.

It was, however, from a railway laboratory (in Derby) that one of the first systematic works on lubrication appeared. This was a joint book by L. Archbutt and R. M. Deeley, a chemist and a locomotive superintendent of the Midland Railway, respectively.[9] From then on, chemistry and railways were indissolubly linked, at least as far as lubrication was concerned.

2.4 WATER

A concern surrounding the operation of *Rocket* was the quality of water used to generate the steam. The directors of the Liverpool and Manchester Railway were anxious to ensure that the purest water was used. Impurities would corrode the iron parts of the boiler. They may also have been aware of the dangers of hard water, where deposited lime created difficulties in cleaning and adherence to the outside of the tubes lowered their steam-raising ability; in extreme cases this could even help to trigger a boiler explosion.

Their concern was in contrast with a common practice soon developed in the USA, where trains on long journeys would sometimes stop to obtain water from nearby ponds or streams by using a water-proof bag to "jerk" it into the tenders. The railwaymen employed for this task were called "jerkers", and many of the railways in the mid-West became known as "jerkwater lines". For the relatively short English line between Manchester and Liverpool, arrangements were made with the Manchester and Salford Water Works, and their product proved apparently satisfactory.

As the century progressed, a large number of steam boilers exploded, both from stationary engines and from those in ships and railway engines. In 1859, over 50 such explosions were reported, and by 1893, that annual number had risen to nearly 100 (though these were not all railway locomotives). Much research was done, and a variety of causes established.[10] Explosions were often due, it was thought, to the use of "hard" water, containing dissolved salts, like calcium bicarbonate. However, some of these disasters were later traced to defective boiler engineering or to the use of too high a pressure (by tying the safety-valves down, for example). On the railways, one of the first recorded cases of a boiler explosion was in 1840, when a Birmingham and Gloucester locomotive exploded at Bromsgrove, at the foot of the notorious Lickey Incline. Both of the footplate crew were killed, and it was impossible to tell whether there had been poor manufacture or inspection, whether the safety-valves had been fixed down in preparation for the heavy work ahead, or whether the chemical action of water was responsible.[11] In any event, when hard water was used, large amounts of solid detritus were left behind (chiefly mineral carbonates) and this posed, at very least, an additional challenge to those who cleaned the boilers after a run. For whatever reason, chemical analysis of the water supplies became essential and was an early feature of railway operation. Many other factors were identified, as in "furrowing", which was largely an engineering defect.[12]

Locomotive tanks were filled with water at standing columns at stations, and later water was picked up by moving trains from water troughs, the first being on the down line of Mochdre, west of Colwyn Bay, an invention of Jas. Ramsbottom, locomotive superintendent of London and North West Railway (LNWR), in October 1860.

Early in railway history, water, however it was delivered, became the first substance for which railway companies obtained analyses (using external consultants). The supply was monitored by chemical analysis to ensure an acceptable quality. This analysis of water became one of the most visible signs of chemical activity on the railways. Eventually many locomotive depots routinely treated boiler water with "lime/soda" – a mixture of hydrated lime (calcium hydroxide) and soda. Even then, a watchful eye had to be kept by chemists to ensure that the water continued to be satisfactorily softened.

2.5 FUEL

An environmentally pleasing stipulation laid down by the organisers of the Rainhill Trials was that each competing locomotive should "consume its own smoke". This was a consequence of an enabling Act of Parliament. The simplest way of complying was for competitors to fire their engines by coke rather than coal. This became increasingly available as a by-product from the making of coal gas, as we have seen. Coal, though an immensely complicated substance, was another mineral analysed by early consultants. Sometimes coke was made by the railways, and others, in specially constructed coke-ovens (which, of course, concentrated the air pollution in one place). Coke was largely pure carbon, though various impurities from the original coal may have been present. The chief benefit of using coke was that no smoke was emitted to create the kind of social nuisance that the promoters feared. With the passage of time, coke became more expensive, due to the competing demands for it by the steel industry. Modern steam locomotives have used coal, with concerted efforts being made by the footplate crew to avoid smoke pollution in heavily populated districts. Developments on the Midland Railway in about 1859 included a redesigned firebox, with a brick fire arch and a deflection plate. This design allowed for better draughting, and hence more efficient combustion, so that, in the hands of an efficient fireman, Midland locomotives emitted minimal smoke.

In running of the railways, coal was consumed in huge quantities, so chemical processes were at the heart of locomotive operation, as in all combustion processes. As a desirable commodity, coal was abundantly available in the northeast of England, where *Rocket* was constructed, in addition to many other places. Modern railways are, after all, direct descendants of the old 18th century colliery lines (even deriving their gauge from some of them: 4′ 8½″). The association between coal and rail is a long one. In Britain, this has meant the supremacy of "King Coal" on the railways, and locomotives burning wood or oil have always been a rarity here.

The burning of coke, a coal product, solved the problem of visible contamination of the atmosphere. The emission of carbon dioxide (from coke or coal) was of little concern, not only at the beginning of railway development but even much later when the gas was identified as a major contributor to the greenhouse effect and subsequent climate change. The impact of carbon dioxide on climate was only acknowledged many years after the period with which we are concerned, but does illustrate a perennial refusal by society to consider things chemical unless necessity demands. We now know that diesel locomotives, and electric trains that depend on coal-powered generating stations, are just as responsible for this effect as the steam locomotive. Chemical measurement of CO_2 emitted into the atmosphere was generally not an issue that demanded chemical action until quite recently (though it was occasionally measured in tunnels in the very early days, chiefly to reassure the travelling public – see Chapter 5).

Locomotive performance was monitored from the very beginning. As early as 1818, George Stephenson was measuring the pull exerted by a colliery locomotive – there must have been many people who needed convincing that it could do better than a horse! The famous Rainhill Trials in 1829 compared the relative effectiveness of three locomotives, and similar comparative tests were performed on many subsequent occasions. Nowadays we can measure the efficiency of a locomotive, and we know that *Rocket* had an efficiency of only 7%. With cheap coke this was hardly an issue, but even today, with expensive coal and "green" issues to address, the improvement is not remarkable: efficiency is only 28% for electric traction, 33% for diesel electrics, 37% for diesel hydraulics and 35% for gas turbine propulsion.[13] At the end of the period studied in this chapter (up until 1923), the *Caerphilly Castle* of the GWR (once *Rocket's* neighbour at the Science Museum and now at Swindon's Steam Railway Museum) had an overall thermal efficiency of only 8.22%, but it could manage far higher speeds and pull heavier loads than *Rocket*.[14] Many of these improvements occurred because the railways of Britain were no longer without chemists. This happened because the engineers realised that the composition of many of their materials could not be left to chance and therefore needed chemical analysis, but also because by the mid-19th century, chemistry itself was poised to advance into many fields of human endeavour, not least the railways. How this desirable state of affairs was reached we shall see in the following chapters.

REFERENCES

1. For example: C. E. Stretton, *The Development of the Locomotive, A Popular History 1803–1896*, Lockwood, London, 1896; republished by Bracken, London, 1989.
2. Frequent references in his private papers: *e.g.* R. B. Wilson, *Sir Daniel Gooch: Memoirs and Diary*, David & Charles, Newton Abbot, 1972, pp. 70, 231, 244 *etc*.
3. (a) M. R. Bailey and J. P. Glithero, in *Perspectives on Industrial Archaeology*, ed. N. Cossons, Science Museum, London, 2000, p. 163; (b) *idem, The Engineering and History of Rocket: a Survey Report*, National Railway Museum, York, 2000; reprinted 2001.
4. J. A. Haswell, in *The Stephenson Centenary*, ed. W. Duncan, E. W. Allen, London, 1881; reprinted Frank Graham, Newcastle-upon-Tyne, 1975, p. 71.
5. N. McCord, in *Literary and Philosophical Society of Newcastle-upon-Tyne, Bicentenary Lectures, 1993*, Literary and Philosophical Society of Newcastle-upon-Tyne, 1994, ch. 4.
6. R. W. Cahn, *The Coming of Materials Science*, Pergamon, London, 2001.
7. G. E. Brown, *Railway Magazine*, 1898, **2**, 53.
8. M. R. Bailey and J. P. Glithero (ref. 3b), p. 29.
9. L. Archbutt and R. M. Deeley, *Lubrication and Lubricants: A Treatise on the Theory and Practice of Lubrication*, Griffin, London, 1st edn, 1899; 5th edn, 1927.
10. See C. A. Smith, *The History of Corrosion Technology*, PhD thesis, The Open University, 1975.

11. L. T. C. Rolt, *Red for Danger*, revised edn, Pan, London, 1966, p. 69.
12. (a) H. Tyler, *The Engineer*, 1862, **14**, 293; (b) H. Tyler, *The Engineer*, 1863, **15**, 311.
13. B. Beer, *Railway Magazine*, 2008 (May), **154**(1285), 12.
14. E. L. Ahrons, *The British Steam Railway Locomotive, from 1825 to 1925*, Ian Allan, London, 2nd reprint, 1963, p. 356.

CHAPTER 3

Chemistry: A New Force in Society and in Railway Development

Most railwaymen and railway historians will see the history of chemistry as a highly specialised subject, fascinating to its adherents, but incredibly boring to everyone else and of little relevance to their own work. They could hardly be more wrong. The general interest of chemical history depends entirely on how it is presented, and the rest of this book is a standing affirmation of its relevance to railways.

3.1 THE FIRST CHEMICAL REVOLUTION: COAL, IRON AND THE REST

The first years of the rise and growth of railways were accompanied by something which some people have called "the chemical revolution". This chemical revolution was involved with the conquest of matter, just as early railways came to be seen as attempts to bring about the conquest of space. This was a revolution in chemical technology, popularised by Archie and Nan Clow in a major book.[1] At that time the Industrial Revolution was in its infancy, and in their outstanding treatment of the "chemical revolution", it is significant that Archie and Nan Clow were

Early Railway Chemistry and its Legacy
By Colin A. Russell and John A. Hudson
© Colin A. Russell and John A. Hudson 2012
Published by the Royal Society of Chemistry, www.rsc.org

obliged to stop at around 1830, the very year in which public railways began. That kind of chemistry had done its job.

Among the substances to be explored in the 18th and 19th centuries were common salt, chlorine, soda and sulfuric acid. Each of these substances had many social implications. The work on chlorine led to the possibility of bleaching the textiles that were flooding the British market in the 19th century; from salt was obtained not only chlorine but also hydrochloric acid (an important chemical reagent); and soda was also used to make soap, glass and much else. Railway carriages had textiles to cover the seats and for curtains, glass for the windows and even soap in the toilets, though not all passengers enjoyed these facilities. Fewer still will have reflected on their chemical origins! However, these trivial examples are not what are meant by the influence of chemistry on railways.

Two other substances had a far bigger role to play: iron and coal. For railways these materials had always been important. It was, after all, the carriage of coal that started off the construction of the railways. The 18th century tramways serving the mines in the north east were created to transport the newly won coal to ships on the Tyne and elsewhere, for distribution (particularly) to London and the south east, as well as overseas. These primitive precursors were truly the beginnings of Britain's railway system, and clearly the primitive "trains" of coal trucks contained iron in some of its forms, although the main structural material was often wood.

First, then, iron. The relevance of different forms of iron for early engines like *Rocket* has been briefly touched on in the previous chapter, but now we must explore this theme rather more fully because of its huge importance for railways as a whole. Iron, present in great abundance in the earth in the form of its oxides, had long been needed for naval ships and for armaments, usually by removing oxygen (by reduction) by means of charcoal made from entire forests of trees. By the 18th century, after many wars, the demand for iron dramatically accelerated and military needs were reinforced by those of agriculture and even domestic tools. Such was the rate of deforestation in parts of England that the process had become a cause of great concern; for example, iron-making in the forests of Sussex came to an end in 1827. Ashdown Forest and many other places in the Weald of Kent, Surrey and Sussex are now renowned chiefly for their leisure potential. The "hammer-ponds of Sussex" were once sources of water power to operate the hammers in the forges, but have now, along with villages like Abinger Hammer, become either folk memories or modern sought-after tourist objectives.

Alternative methods to simple charcoal reductions had been sought for some time, including the injection of blasts of hot air through the hot mixture. However, the first major advance, at least in England, came in the Shropshire town known as Coalbrookdale, quite justifiably known as "the cradle of the Industrial Revolution". At the beginning of the 18th century, a family led by a Quaker foundry man, Abraham Darby (1678–1717), settled in the area.[2] At first Darby used charcoal for reducing iron ore, but he soon moved over to coke, taking advantage of local deposits of coal. He was lucky in finding a suitable kind of coke, and tried this (instead of the customary charcoal) to get iron from its ore.

That was not all. Soon after this, Abraham Darby began to employ coke that had been made on a small scale for the local malting business, but soon produced his own by heating coal in the relative absence of air. The immediate product from his furnaces was "cast iron", a brittle form of the metal that contains up to 4.5% of carbon (see Figures 3.1 and 3.2). Cast iron can be melted easily and flows into moulds, being strong in compression but weak in tension. Removal of much of the carbon by a variety of methods

Figure 3.1 Abraham Darby's iron furnace at Coalbrookdale (the roof has been added much later to aid conservation).

Figure 3.2 Early piece of flanged cast-iron waggonway, probably from the 1760s or later preserved at Coalbrookdale.

Figure 3.3 Ironbridge. The first bridge in the world made of iron. With permission from Ogilvy & Mather, New York.

gave "wrought iron" (which had lower carbon content, and was strong in tension but weak under compression). Much later steels were obtained with intermediate carbon contents, and often with additional trace metals added.

By 1709, Darby had succeeded spectacularly and was able to make large quantities of "pig iron", as it was called. This was what he needed for his foundry. His grandson, Abraham Darby III, was able to make the world's first iron bridge (to cross the Severn Gorge) in 1777–9, and Ironbridge (as it is still called) has given its name to the whole site (see Figure 3.3).

Other iron-making works followed Darby's lead, the first one in Scotland being established at Carron in 1759. The production of wrought iron and, much later, steel involved more new technologies, and the railway industry of Britain, with its unceasing demand for all kinds of iron, became one its chief beneficiaries.

The conversion of coal to coke also intrigued an eccentric and impecunious nobleman in Scotland. He was Archibald Cochrane, 9th Earl of Dundonald,[3] who in 1778 had inherited an estate in the picture-postcard village of Culross, on the north bank of the Forth, a little upstream of the present Forth Bridges. Finding his estate was rich in coal he decided to experiment with it. From his grounds, almost out of the village, and on a small hill, he tried the effect of heating large quantities of coal in almost closed vessels (see Figures 3.4 and 3.5). He obtained what he wanted, coke in abundance, and also large quantities of tar. As well as these, he produced masses of a gas (coal gas) which burned with such an intense flame that it could be seen from the opposite bank. Thereafter this was continued for its entertainment value, but Dundonald had discovered not only a method for making coke, but also vast amounts of tarry liquid and this inflammable gas.

Figure 3.4 Culross House, where Dundonald made his classic experiments on coal distillation (on the lawns that slope down to the Forth).

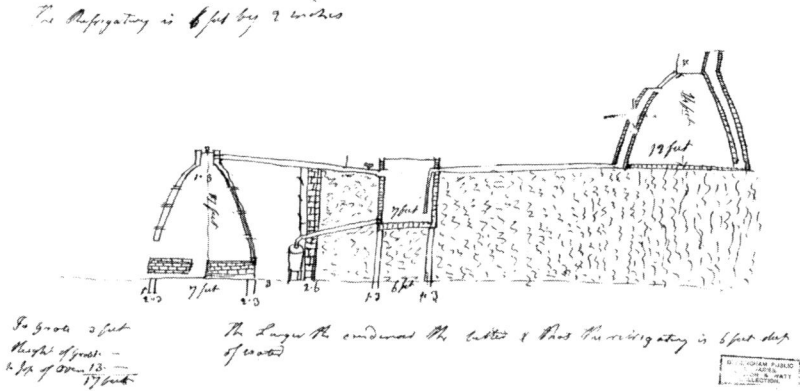

Figure 3.5 Lord Dundonald's experiments. Reproduced with the permission of Birmingham Libraries & Archives.

Before long the gas lighting of houses and streets was being introduced, and this of course included the premises of the infant railway industry. As for the tar, though at first a nuisance, it was found that, following distillation, the product could be used as a wood preservative and it was much used on fences and, in particular, on wooden railway sleepers. The exploitation of coal tar for chemical purposes came later.

In this account of the first chemical revolution the emphasis has been on a poorly understood exploitation of inanimate matter. Then, towards the end of the 18th century another, second, chemical revolution was struggling to emerge. Whereas the first revolution was concerned with the *manipulation* of matter, the second was more about *understanding* it. A technology had blossomed into a science. It was not that up until now there had been one erroneous idea about the nature of substances, but rather that there had been many. Now, at last, modern chemistry was being born.

3.2 THE SECOND CHEMICAL REVOLUTION: ATOMS AND MOLECULES

Chemistry is about the science of all the substances that may be encountered in the universe. Some of these are incredibly complicated, so one task for the chemist is to understand how they are constructed. Fortunately, the variety of their basic units is very small, and these are known as chemical "elements". It was just about at the dawn of the railway age that the nature of these elements became clear, and chemistry at last became something like the science we recognise today. So several thousand years of human history had passed before, less than two centuries ago, chemistry came of age. The history of chemistry has often been discussed[4] and much has been published on the individual phases of development of different parts of the theory. We shall focus on just a few basic parts of chemistry's theory and practice, however, and see how it fitted into Victorian society and became a topic the railways did not dare to neglect.

To understand how a complex machine like the *Rocket* works, one has first to take it to pieces, literally or figuratively. You need to know about its boiler, cylinders, chimney, wheels, brakes, and so on. You have to look carefully at each component and then see how they all fit together. In chemistry, unlike engineering, the individual components of substances are often quite different from their parent substances!

Although most substances presented to us by nature are much more complicated than the *Rocket* (such as Vitamin C), many are much simpler. When all substances are analysed, we find that the basic components are things called chemical elements, of which only a couple of dozen are very common. Elements are like the Lego bricks assembled to make a model house, church or whatever. These elements cannot be broken down into anything simpler and still be material substances.

Two substances that were for years thought of as elements are vital to a steam engine: air and water. We now know that they are not elements at all. Air is a mixture of mainly two gases that are indeed elements: nitrogen and oxygen (with a few other gases in much smaller amounts). The components of air can be separated by mechanical means, such as by strong cooling followed by distillation. Water, however, is also composed of two elements, but these are chemically combined and cannot be separated anything like so easily, and when this does happen the sum of their properties is nothing like the water from which they came. The elements from which water is made are oxygen and hydrogen. We will now examine how this knowledge about elements came about.

Through the centuries men had pondered the question as to what substances were made from, and various chemists got very close to the answer, including the English chemist, Joseph Priestley, and the

Swedish chemist, Carl Wilhelm Scheele, both in the late 18th century. It fell to the French chemist, Antoine Lavoisier, to make the decisive discovery and the clearest declaration. Unfortunately he was guillotined during the French Revolution, though that had nothing to do with his chemistry (this is one of the many human stories abounding in the history of chemistry but we cannot pursue it here).

Up to around 1800, chemists had been fascinated by that most common of chemical reactions: fire or combustion. When they burned wood (for example) it seemed as though something was being given off into the atmosphere. This was called "phlogiston". When, in the later 18th century, chemists began to weigh (for instance) a metal and the solid it left behind after combustion, they found that the product was heavier than the metal. In giving up its phlogiston, as they supposed, it had gained in weight, and so phlogiston must have a negative weight! What Lavoisier and others were to show was that combustion did not involve the liberation of anything like phlogiston, but the combination of the burning substance with an element of the air which they now called "oxygen". Many, like Priestley, hung on as long as they could to the phlogiston theory, but gradually the number of elements was extended to include oxygen. Lavoisier was able to publish a list of elements that seems astonishingly modern, even though a few interlopers still remained. As the railway age dawned, chemists at last knew the units of their substances, the chemical elements. And that was a great help. Much more was to come.

If everything can be seen as composed of elements, what are the ultimate particles like? For centuries it was thought that if you divided a thing and continued to do so as far as you could get, you would end up with particles that were unsplittable, or atoms.

At the beginning if the 19th century, a Quaker then working in Manchester, the Cumbrian John Dalton (Figure 3.6), proposed his atomic theory (see Figure 3.7). Not only was matter composed of these tiny particles of these elements, known as "atoms", but, *atoms of each element all had the same weight, the atomic weight.*[5] This was a tremendous breakthrough, and made sense of the many reactions where components (and often product) had a simple weight relationship to one another. Thus 1 part by weight of hydrogen combined with 8 parts by weight of oxygen to produce 9 parts of water.

Figure 3.6 John Dalton (1766–1844).

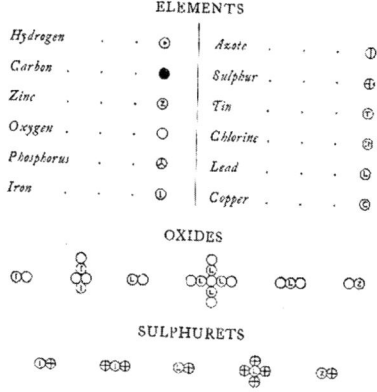

Figure 3.7 Dalton's atoms.

The problem was to know what that weight was! Fortunately some of the simplest elements are gases, like oxygen, nitrogen and hydrogen. They were given simple symbols, such as O, N and H. By taking equal weights of hydrogen, and then weighing the oxygen, nitrogen or whatever element that combined with this weight, he could tell their combining weights relative to hydrogen = 1. So oxygen became 8, and nitrogen became approximately 4.6. These were assumed to be their atomic weights. From these he calculated the atomic weights of other elements, for example, the weight of copper combining with 8 parts of oxygen.

Sadly, Dalton had assumed that each gas was monatomic – having discrete single atoms of hydrogen or oxygen wandering about without being joined to anything else. He also assumed that water had a formula of HO rather than H_2O. So, on the basis of their combination to form water, he thought their atomic weights were in the ratio 1 : 8. We now know he was wrong, and also the usual pattern for hydrogen and oxygen is that their ultimate particles (molecules) consist of *two* atoms joined together, in other words they are H_2 and O_2. Similar conclusions were eventually reached for nitrogen and chlorine. By 1830 many, though not all, chemists accepted Dalton's atomic view of matter.

But should the atomic weight of oxygen be 16, not 8? The clue to this problem was to lie in the volumes, not the weights, of the gases involved, though Dalton would have none of this. The dilemma for chemistry persisted right up to the 1860s, though the answer was embedded in an article by an obscure Italian scientist, Amedeo Avogadro, back in 1811. His famous "hypothesis" proposed that equal volumes of all gases contain equal numbers of *molecules*, not *atoms*. These molecules, like O_2 and N_2 were the normal state for the gases involved, but they could still be broken down into their component atoms. Further studies on solids enabled a picture to emerge of the atomic weights of many elements relative to oxygen, but who was to say that oxygen gas was O_2 rather than O or even O_3?

A resurrection of Avogadro's hypothesis by a fellow-countryman named Stanislao Cannizzaro, led to a resolution of this difficulty and confirmed the atomic weight of oxygen as 16, not 8. Many other figures had to be revised as well. Unfortunately the community of European chemists could not agree, and eventually (as we have seen) they convened the Karlsruhe Congress in 1860, the first international conference of chemists. In the end it became clear that Cannizzaro was right all along. This meant that now chemists could agree on formulae like CO_2 and H_2SO_4, and many much more complicated molecules. Knowing the composition of compounds did not explain how the atoms were held together, however, and indeed how many of them were present. Why, for example, was sulfuric acid H_2SO_4 and not H_4SO_2? To answer these questions a further concept had to be introduced: valency.

The concept of valency (or valence in the USA) did not emerge in a neat and predictable fashion, any more than most of the other chemical theories did.[6] In a sense it all started with the work of the young Lancastrian, Edward Frankland (see Chapter 1). Working in London in the early 1850s on a very obscure range of compounds that he had discovered, Frankland gradually recognised regularity in the formulae of related compounds and concluded that each atom had a definite combining power. This was eventually called the atom's "valency", and was manifest in a certain number of "bonds" to other atoms. Thus the simplest organic (carbon-containing) compound was methane. Known to be CH_4, it was more fully written as shown in Figure 3.8, where the straight lines indicate the *bonds* linking the atoms: carbon has a valency of 4 and hydrogen 1. Later work showed that the molecule is in reality not planar, as shown here, but a three-dimensional affair, though we usually show it as a planar projection. The famous hexagon formula (and assumed structure) for benzene came from August Kekulé in around 1865. Thus every molecule had a structure unique to itself, and when (for instance) railway chemists had problems relating to lubrication, the structure of each component of the oil could, at least in principle, be written down and understood.

One other development, at that time, was a discovery by the Russian chemist Dmitri Ivanovich Mendeléef, and (independently) by the German, Lothar Meyer, that chemical elements can be arranged into what was called the "Periodic Table", and that many elements fell into well-recognisable families (1869). Thus halogens (fluorine, chorine, bromine and iodine) are all reactive non-metals with a valency (among others) of 1, while the rare gases, such as helium, neon, argon, krypton and xenon (discovered later), are very unreactive and were thought at one time to have a valency of 0; these were the "inert gases". This area of research was generally considered the province of "inorganic chemistry".

During the 19th century there was great progress with "organic chemistry" – the chemistry of carbon atoms. Carbon has the

Figure 3.8 Methane.

unique tendency of forming stable bonds with itself, thus benzene, C_6H_6, has its six carbon atoms arranged in a ring, and so have thousands of its derivatives. The number of possible organic compounds is immense, and vast numbers of synthetic drugs and dyes began to be made, particularly in Germany. At this time a third chemical discipline emerged, later called "physical chemistry", dealing with such diverse subjects as electrolytic cells, diffusion, osmosis, heats of reaction and other subjects directly relating to the operation of the railways. None, however, was as important as analysis.

Around 1800, several workers (such as the Swedish chemist, Jöns Jacob Berzelius; Figure 3.9) achieved the most astonishing accuracy in the analysis of inorganic materials, and the results can be used to give atomic weights of metals that are very near the modern values. Others, supremely Liebig at Giessen (Figure 3.10), developed their own methods of organic analysis and were visited from all over the world.

Some volumetric analysis (measurement of liquid volumes) was also carried out, especially in the late 18th century, but the familiar burette did not appear till around 1850. Gas analysis was tried after

Figure 3.9 Jöns Jacob Berzelius (1779–1848).

Figure 3.10 J. von Liebig (1803–1873).

gases had been recognised from the early 18th century, and was later perfected by Bunsen and others. Some qualitative tests familiar to chemists till very recently came from the 18th century, including the charcoal block test, the borax bead test and the brown ring test. Many more were invented in the years that followed.[7]

It was the 19th century that saw incredible growth in the practice of analysis. Partly this reflected the availability of cheap glass instruments, partly the work of pioneers like Berthollet and Mohr, and partly a new realisation of its social utility. Thus analysis was employed in the detection of poisons,[8] especially arsenic,[9] and in the testing of proprietary medicines.[10] But there was another factor, too. Chemical analysis, or analytical chemistry, has a long history.[11] It is generally agreed that it began in the Ancient World, with the assaying of metals. For this process weighing was all-important, and so the chemical balance has a long and respectable history.[12] The chemical balance is still used in railway chemistry, and is a prominent feature of old photographs of railway laboratories. By 1830, chemists had much experience of analysing mineral waters to identify the beneficial and the harmful components, an experience which made them very useful to railway companies at the start of the railway age.

Alongside the colliery lines and the passenger railways, other industries were growing. The new textile establishments required dyestuffs of guaranteed composition in order to achieve satisfactory colours. More significantly, the rising heavy chemical industry required analysts. The quality of starting materials was often of the greatest importance if trade was to succeed. Analyses of their products had to compare favourably with those of rival firms, or trade would be affected. From 1863, an Alkali Act ordained that the gaseous effluent from acid plants should be monitored rigorously in order to minimise air pollution. For all these functions chemists were required. It may be significant that in the early days of railways, many chemical laboratories existed near the River Tyne, an area abounding in all kinds of chemical industries (as in Figure 1.2). The expertise of resident chemists might be called on for railway purposes, as was the case of Thomas Richardson (see Chapter 4).

During these years, independent analytical chemists were setting up their own practices, such as Richardson (1844) and J. Pattinson (1860) both from Newcastle, as well as many others in London. These were some of the chemists who, as consultants, assisted the railways in their early days. The existence of chemically adulterated food led to two Acts, in 1860 and 1872, that permitted and then demanded a new set of chemists, the Public Analysts.

Thus by the middle of the 19th century analytical chemistry had come into its own. At the hands of Berzelius, Liebig and others, atomic and molecular weights had been determined by analysis. The rapid development of technical methods of analysis in general proved valuable for all kinds of railway purposes.

3.3 A SCIENTIFIC SOCIAL REVOLUTION: ENTER THE CHEMIST!

Through the ages men (and some women) had been known for their prowess in manipulating matter; they had been called by various names (some distinctly uncomplimentary), and others as "alchemists" or "adepts", but rarely "chemists". There was then no recognisable science of chemistry, though the name could be derived from the untidy and complex mixture of ideas and manipulations which some people knew as "alchemy".

If "railways without chemists" (at least on their permanent payroll) marked their first few decades, then "chemistry without railways" was the situation before 1830. It is interesting that the year of the Rainhill Trials saw also the demise of the most colourful showman and chemical investigator of Britain, Sir Humphry Davy.[13] He had not only invented a safety lamp for coal mines, independently of George Stephenson, but had also been the first man to publicise chemistry in his lectures at London's new Royal Institution and to make it a topic of lively conversation amongst those who could afford the time and money to attend. His most important contribution to science, however, was to establish new links between chemistry and electricity. He did not discover the electric battery or the phenomenon of electrolysis. However, he certainly used them to spectacular effect and was able to isolate the highly reactive elements of potassium and sodium for the first time, together with some less reactive elements. Later Davy proposed an electrochemical theory, where chemical forces were seen as fundamentally electrical in nature. Thus ideas were developed in a way that was to dominate chemical thinking such as by the Swedish chemist, J. J. Berzelius.

By 1830, Davy was dead and chemistry has lost a brilliant spokesman. Others, especially in London, continued to sell the subject to the poorer classes, and a small but steady supply of such men, nearly all without formal training, began to appear. Some worked in the up and coming factories, and a few designed new dyes for the textiles (*e.g.* calico) that were in much demand. In 1825 the Government withdrew the Salt Tax. Since this material was a starting point for the manufacture of many large-scale chemicals, the chemical industry began to grow, and with it a demand for chemists. The price of salt dropped to one sixth of its previous value within 20 years. In 1831, two interesting events took place. One was the repeal of a tax on printed calico, and dyestuff chemists soon began to prosper; the other was the appointment of the first brewery chemist, Robert Warington to Truman, Hanbury, Buxton & Co. Was no industry exempt from chemical assistance? Probably not, if it employed recognisable chemicals.[14] The tax on glass was repealed in 1845, and great factories could now be built with glass panes to let in light and exclude cold air. Again, more chemists were needed in factories that made soda or converted it into glass.

Training of chemists, at the Royal College of Chemistry in London (from 1845; see Figure 3.11), the Mechanics' Institutes, and elsewhere was one of the factors taken into account before a budding "chemist" could be admitted into the profession. Now potential employers, the railways included, could at last recognise a qualified chemist.

While the second chemical revolution was gaining momentum, the academic chemists of Britain realised that they had much in common. They created for themselves what became the world-famous Chemical Society of London (1841; see Figure 3.12), the first such national institution anywhere in the world. Other countries were quick to follow Britain's example. By publishing papers, convening discussion meetings, and even by establishing a headquarters in London with its own library and meeting rooms, they encouraged the growth of chemical understanding on a scale that had never been seen before. In this respect they even eclipsed the more famous Royal Society of London, whose prestigious activities included all the other natural sciences, but did not descend to the level of detail in some chemical matters that the railways were to

Figure 3.11 Royal College of Chemistry, Jermyn Street, London. With permission from College Archives, Imperial College, London.

Figure 3.12 The Royal Society of Chemistry (previously the Chemical Society of London), Burlington House, Piccadilly, London. With permission from the Royal Academy of Arts, London.

demand. Some of the leading Fellows of the Chemical Society were, however, also honoured with Fellowship of the Royal Society (FRS).

Other, less prestigious chemical groups came and went, but there was nothing to deal with the now pressing and related problems of status, remuneration and recognition by Government. It is a long story, made more complicated by the fact that in English (alone) the same word *chemist* is often used by practitioners of chemical science on one hand, and by pharmacists and druggists on the other. Much agitation and complaining led eventually to the foundation of the Institute of Chemistry in 1877.[15] Its first President was Edward Frankland. This was the world's first *scientific* professional organisation, although the mechanical and civil engineers had enjoyed their professional institutions since earlier in the 19th century.

The rest of this book will show how indebted to chemistry the railways were to become. Nevertheless, it is appropriate to point out that the debt was also sometimes the other way round. Just as the railways had established the pattern of a seaside holiday for the masses, so also they helped to foster the practice of national chemical conferences. This was obvious within Britain where the Chemical Society had regular congresses to which travel was only possible by train. Much later on, in the 1930s, the Society, faced with a declining membership, promoted provincial conferences of which 29 took place in 1934. Again, railway travel must have been highly significant. But now the public railways of the UK were beginning to redress the balance of debt by showing how important to them chemistry had become.

REFERENCES

1. A. and N. Clow, *The Chemical Revolution: A Contribution to Social Technology*, Batchworth Press, London, 1952.
2. See also the 2003 Rolt Memorial Lecture, D. de Haan, *Ind. Arch. Rev.*, 2004, **26**, 3, with full references to other sources.
3. These and many related issues are further explored in C. A. Russell (ed.), *Chemistry, Society and Environment: A New History of the British Chemical Industry*, Royal Society of Chemistry, Cambridge, 2000.
4. A straightforward account of the whole story has been given by one of the present authors: J. A. Hudson, *The History of Chemistry*, Macmillan, London, 1992. More detailed works include W. H. Brock, *The Fontana History of Chemistry*, HarperCollins, London, 1992, and several others; details of many of the ideas may be found in C. A. Russell, *From Atoms to Molecules: Studies in the History of Chemistry from the 19th Century*, Ashgate, Farnham, Feb. 2010.
5. Most modern science recognises that some atoms may be split (as in radioactivity), and that not all atoms of the same element have exactly the same weight; there are such things as isotopes, and modern "atomic weights" are averages, taking these into account. But for ordinary purposes, then as now, most of the basic concepts of Dalton remain true.
6. C. A. Russell, *The History of Valency*, Leicester University Press, Leicester, 1971.
7. For a more specialised account of some of these, and other, methods see F. L. Holmes and T. H. Levere (ed.), *Instruments and Experimentation in the History of Chemistry*, MIT Press, Cambridge, MA, 1999.
8. W. A. Campbell, *Anal. Proc.*, 1980, **17**, 76.
9. W. A. Campbell, *Chem. Br.*, 1965, **1**, 198.
10. W. A. Campbell, *Isis*, 1978, **69**, 226.
11. See F. Szabadvary (translation B. Svehla), *History of Analytical Chemistry*, Pergamon Press, Oxford, 1966; *idem, Periodica*

Polytechnica (Budapest) 1977, **21** (no. 4), 355; D. Betteridge, *Anal. Chem.*, 1976, **48**, 1034.
12. J. T. Stock, *Development of the Chemical Balance*, HMSO, London, 1969.
13. There are many book-length biographies of Davy. Perhaps the most useful for historians of technology is still Sir Harold Hartley, *Humphry Davy*, Nelson, London, 1966 and S. R. Publishers, Wakefield, 1971; see also D. Knight, *Humphry Davy: Science and Power*, Blackwell, Oxford, 1992; R. Lamont-Brown, *Humphry Davy: Life Beyond the Lamp*, Sutton Publishing, Stroud, 2004; and several others.
14. On the chemical industry see especially ref. 1 and 3.
15. C. A. Russell, N. G. Coley and G. K. Roberts, *Chemists by Profession*, Royal Institute of Chemistry/Open University Press, Milton Keynes, 1977.

CHAPTER 4

Testing the Waters: The Early Railways and their First Chemical Consultants

4.1 THE DAWN OF THE RAILWAY AGE

September 15th, 1830 was a date of enormous importance in railway history, for on that day the line between Liverpool and Manchester was opened. Other railways had preceded it, most notably the Stockton and Darlington in 1825. However, there were a number of features of the new line which made it special, and the most important of these was that it was the first to rely entirely on mechanical traction. Although the horse was to be a beast of burden for many years to come, some of the more far-sighted onlookers probably realised that horses would ultimately be replaced by man-made machines.

The opening day was one of great celebration (see Figure 4.1), and was the culmination of a lengthy period of preparation. The line itself had taken four years to construct, and the choice of locomotive had been made at the famous Rainhill Trials in 1829, when *Rocket* demonstrated its superiority over its competitors.

Early Railway Chemistry and its Legacy
By Colin A. Russell and John A. Hudson
© Colin A. Russell and John A. Hudson 2012
Published by the Royal Society of Chemistry, www.rsc.org

Figure 4.1 The scene at the opening of the Liverpool and Manchester Railway, September 15th, 1830. (Science Museum/SSPL).

4.2 THE PROBLEM OF BOILER SCALE

One of the great technical advances incorporated into *Rocket* was the multi-tubed boiler. The hot gases from the firebox were conducted through tubes in the boiler to the smokebox at the front of the locomotive, from where they emerged through the chimney. This arrangement provided a much larger surface area for heating the water than in earlier more primitive locomotives. Steam could now be generated at a much faster rate than previously, and hence a locomotive could travel for extended periods without running out of steam and stopping. However, if hard water is used in any steam boiler, the heated surfaces become covered with a layer of scale, which acts as an insulating layer, decreasing the efficiency of the boiler. The most important chemicals present in scale are calcium carbonate and calcium sulfate, which in the nineteenth century were called carbonate of lime and sulphate of lime, respectively. These chemicals come from the passage of the water through calcium-containing rocks: the calcium sulfate is simply dissolved, while calcium carbonate (as in chalk and limestone) is converted by dissolved carbon dioxide into the soluble bicarbonate which reverts back to the insoluble carbonate when heated.

Removing the scale from boilers had been a problem ever since the introduction of stationary steam engines to pump water from mines, and by 1830 chemical consultants were being asked to advise on the suitability of water supplies for the boilers of the many steam engines then in use. In that year, one such consultant, William West, published a paper in which he discussed the chemical nature of the scale formed in the boilers of stationary steam engines. While it had previously been known that carbonate of lime was a component of boiler scale, West pointed out that sulfate of lime is frequently present as well. He described how he had been informed at Manchester that "it was necessary to empty the engine boilers, and chip out the crust formed, in some cases as often as six weeks". West, one of the many Quakers at the forefront of the Industrial Revolution, noted that a further disadvantage of the deposits was that the Sabbath was being employed for the descaling operation.[1]

Scale in a multi-tubed locomotive boiler was much harder to remove than scale in a boiler of a stationary engine. In the case of the latter, a workman could crawl inside the empty boiler to chip out the scale, but in the case of a locomotive boiler this was of course impossible, and if a thick layer of scale formed, the boiler might have to be re-tubed. The rapid development of the locomotive was making this potential problem much more serious. While *Rocket* had 25 tubes each of three inches in external diameter, *Northumbrian*, delivered to the Liverpool and Manchester a few weeks before the opening day and used to haul the Duke of Wellington's carriage on that occasion, had 132 tubes each of $1\frac{5}{8}$ inches in diameter. A thick layer of scale around the tubes not only reduced the rate at which steam was generated, but the tubes deteriorated more rapidly because of the higher temperature reached inside. It was also believed that the deposited scale was a cause of railway boiler explosions, as it was thought that if a portion of scale broke away from the outside of a tube, water coming into contact with the red hot surface would lead to the rapid generation of steam and result in an explosion.[2] Certainly boiler explosions occurred with early locomotives, although many were probably caused by failure of the boiler barrel or by inadequate safety valves. Whatever the cause, such accidents were seen by the public as a potential danger of this new form of transport (see Figure 4.2), and the railway companies naturally wished to do

Figure 4.2 Hugh Hughes, "The Pleasures of the Railroad, Shewing the Inconvenience of a Blow-Up" (1831) (Ironbridge Gorge Museum Trust, Elton Collection).

everything they could to minimise the possibility of an explosion. Therefore, for reasons of maximising locomotive efficiency, minimising maintenance costs and reducing the chances of a boiler explosion, finding water supplies which were as pure as possible was a vitally important consideration for a railway company.

4.3 EARLY WATER ANALYSES FOR RAILWAY COMPANIES

Four months before the Liverpool and Manchester line opened, the directors addressed the question of the quality of the water they were considering purchasing. At the board meeting of 10 May 1830, they discussed a proposal from Manchester and Salford Waterworks to supply 10 000 gallons of water per day at a cost of £70 per annum. The directors instructed George Stephenson to report on the suitability of the water for the locomotives. However, Stephenson was either too busy or felt that he was not up to the task, because on 26 July he was asked again, and it was also agreed to write to Dr Henry for his opinion on the quality of this water.[3]

The Dr Henry from whom information was sought was William Henry. Like a number of chemists of the day, he was medically qualified. In 1801, he had published a chemistry textbook which contained a section on the analysis of mineral waters. By 1830 he was a rich man, having made a fortune from the manufacture of magnesia for the treatment of indigestion.[4] No record of any advice provided by Henry to the Liverpool and Manchester Railway appears to have survived, but as a manufacturer he would almost certainly have been familiar with the properties of the local water supply. The verdict on the water must have been favourable, for on 9 August it was agreed to its purchase on the terms offered.[5]

The first chemical analysis for a railway company which is known to have survived was performed five years later by William West (Figure 4.3). This was the William West who, in 1830, had commented on the nature of the scale in the boilers of stationary engines, and he was to become an extremely important figure in advocating the value of water analysis to railway companies. West was born in Wandsworth in 1793 and, after studying pharmacy in London, he established himself in 1816 as a wholesale and retail chemist and druggist in Leeds. He quickly gained a reputation as one of the leading scientists in the locality, and in 1831, he was appointed lecturer in chemistry at the Leeds School of Medicine.

Figure 4.3 William West (1793–1851) (*Chemist and Druggist*, 1960, **174**, 71).

In 1841 West was one of the 77 original members of the Chemical Society (one of the forerunners of the present-day Royal Society of Chemistry). In 1846, he was elected FRS but in the same year he was forced to resign his lectureship owing to ill health. West died in 1851.[6] By 1835, as well as investigating the problems hard water could cause in steam engine boilers, West had also advised on the suitability of water supplies for domestic and therapeutic use. In 1823, he had published an analysis of the mineral water from a new well at Harrogate, followed in 1827 by an analysis of the water from a new spring discovered at Wakefield. The therapeutic value of these waters seems not to have been in doubt; West performed these analyses in order to discover the identity of the beneficial substances they were thought to contain.[7]

The railway company for which West provided his 1835 analysis was the Stockton and Darlington Railway. Presumably the company chose West because of his reputation as a water analyst, although another possible reason is that Edward and Joseph Pease (chief promoter and treasurer of the Stockton and Darlington Railway, respectively) were, like West, Quakers. West analysed four samples of water and his report is of huge importance to the history of railway chemistry, reads as follows:

> I have analysed the four waters sent to me in order to ascertain their fitness for Engine Boilers; the results are No 3 containing 140 grains of solid matter per gallon, chiefly Carbonate of Lime, but with a large proportion also of Sulphate of Lime, the worst constituent usually found in water; this water should be rejected at all cost, as it is utterly unfit for any such use. No 1 contains $24\frac{1}{2}$ grains of solid matter consisting of Sulphate of Lime, Carbonate of Lime and Carbonate of Soda; the proportion of the former is probably small and the latter salt is harmless so that if this water were clear it would be very good, but the sample sent to me is discoloured by some vegetable matter which I fear will render it, while in that state, liable to the defect which our Engineers call "Priming" or frothing so as to drive the water into the steam-pipes.
>
> Nos 2 and 4 are both excellent waters, and resemble each other so nearly that a very precise analysis was necessary to determine exactly their relative values, the results as given below,

shews [sic] that No 2 is the best but that the difference is so slight that any circumstances of abundance, or situation, may be allowed to determine between them.

No 2 contains per Imp Gall, of	Sulphate of Lime	7 Grains
	Carbonate of Lime	5 Grains
	Carbonate of Soda	8 Grains
	Total solid content	20 Grains

Besides a little vegetable matter and traces of Magnesia.

No 4 contained in 1 Gall,	Sulphate of Lime	$11\frac{3}{4}$ gr
	Carbonate of Lime	7 gr
	Carbonate of Soda	$5\frac{3}{4}$ gr
	Total solid content	$24\frac{1}{2}$ grains

With an unimportant proportion of Magnesia.

Leeds 23rd of 9th Month 1835   William West

West apologised for the delay in submitting the analyses, explaining that when the samples arrived he was "in a distant part of the kingdom, on account of his health". He also submitted his account, charging two guineas each for the detailed analyses and 1 guinea each for the other two.

The Stockton and Darlington Railway had opened in 1825 with only one locomotive, the bulk of the haulage being provided by horses or by stationary winding engines. Within a short time, railways were using locomotive power almost exclusively, and the more astute companies, like the Liverpool and Manchester Railway, obtained water analyses prior to the opening of the line. Thus in 1839, the Hull and Selby Railway, which was to open in 1840, commissioned analyses of water from two sources.[9] In 1840, when the Bristol to Bath section of the Great Western Railway opened, the company received two analyses of "Bristol Well Water" performed by a local chemist, William Herapath.[10] Not surprisingly, some companies which did not take the trouble to obtain a water analysis before commencing operations subsequently encountered problems. For example, in April 1838, when the London and Birmingham Railway was still only partially open, Edward Bury, the locomotive contractor, expressed concern about the quality of the water available: "The water at Rugby is so exceeding bad that it is almost impossible to work the line, and it causes delay to the trains daily."[11] In 1842, the Directors of the Newcastle and Carlisle Railway heard that "Many complaints having been made as to the state of the water supplied at Newcastle Station which was found to be injurious to the engines." The secretary was instructed to write to the water company and negotiate a reduction in the charge levied for supplying the water.[12]

4.4 WATER ANALYSIS BECOMES STANDARD PRACTICE

Experiences such as those of the London and Birmingham and the Newcastle and Carlisle Railways may have helped to convince engineers, railway managers, and board members of the importance of obtaining a chemical analysis of an intended water supply. However, railway engineers could have been in no doubt about the merits of water analysis after 1846, when West read a paper at a meeting of the Institution of Civil Engineers. This was an important forum for the discussion of engineering matters, and among those present were Robert Stephenson (of *Rocket* fame), John Viret Gooch (locomotive superintendent of the London and South Western Railway and an elder brother of Daniel Gooch), and Samuel Morton Peto (railway contractor and civil engineer). John Rennie chaired the meeting. West's paper was entitled "On Water for Locomotive Engines and its Chemical Analysis".[13] In it he stated that his object was to point out to engineers "how they may best apply the resources of chemistry".

It is clear that, after his initial work for the Stockton and Darlington, West had built up a considerable practice in performing water analyses for railway companies. In his paper he quoted the results of 16 such analyses, and, as in his earlier work for the Stockton and Darlington Railway, each analysis was accompanied by his opinion as to the suitability of the water for use in locomotives. West noted that "…the engineer, when not acquainted with chemistry, probably takes more notice of the opinion attached by the chemist to his numerical statement, than of the statement itself". Thus the skill and experience of the chemist was involved not only in performing the analysis but in interpreting the results for the benefit of his client.

Echoing the experiences of railway companies such as the London and Birmingham and the Newcastle and Carlisle Railways, West pointed out that there had been a number of occasions when it had been discovered that a water supply was totally unsuitable after the line had been opened. He commented "During the laying out of the line, or the fixing of the site of a station, the quality of the water might form an element of consideration; but it is only when the water is found to be unfit for use that an analysis is thought of". West's advice was timely, as the construction of a large number of lines was about to commence (the second period of "railway mania"). No doubt the publication of the paper also helped to publicise West's analytical services.

West had commented that the result of a water analysis might exert an influence on civil engineering operations during the construction of a line. An example of this taking place comes from a slightly later period, when the South Durham and Lancashire Union Railway was being built (1859–1861). The line was to run from Barnard Castle westwards across the Pennines to Kirkby Stephen and on to Tebay, where it would link up with the Lancaster and Carlisle Railway.

Thomas Richardson was the chemist chosen to perform the analysis of potential water supplies.[14] Richardson (1816–1867) was a native of Newcastle-upon-Tyne, and after studying chemistry in Glasgow, Germany and France he returned to Newcastle in 1838 to set up as an analytical and consulting chemist, taking out a number of patents for chemical processes. In 1848 he became lecturer in chemistry at Newcastle School of Medicine, and from 1856 he was lecturer in chemistry at the University of Durham.[15] Although he came on the scene slightly later than William West, Richardson was, like West, a leading figure in the early application of chemistry to the railway industry. As well as performing water analyses, he also analysed coke and advised on timber preservation (see Chapter 5).

The water supply originally selected by the railway company at Barnard Castle station (from Percy Beck) was apparently judged to be unsuitable. Richardson was asked to analyse possible alternatives. Just after crossing the River Tees at Barnard Castle, the line passed close to an artificial fish pond near the village of Lartington. In August 1860, Richardson analysed a sample of water from "Lartington Pond" (which was also known as Crag

Pond) and found that it contained 3.8 grains of solid matter per gallon, a "remarkably small quantity". On receiving this analysis, William Bouch, the locomotive engineer, forwarded it to the company secretary recommending that he reach an agreement with the owner of the pond before the latter became aware of the excellent properties of the water.[16] The unnamed owner of the pond was the Rev. Thomas Witham, a Roman Catholic priest who owned the Lartington estate,[17] and who was also a director of the South Durham and Lancashire Union Railway.[18] Bouch may have been reluctant to spell out Witham's name, but he had no qualms about suggesting that he should be kept in the dark about the quality of the water. The intention was to pipe the water across the Tees to Barnard Castle station, and it was estimated that if this supply were used, the water would have to be piped a distance of over $3\frac{1}{2}$ miles.[19] There was, however, another fish pond on the estate that had only just been constructed.[20] This was referred to as "Lartington New Pond" or "Low Fish Pond".

Less than two weeks after Richardson had sent the analysis of the water from the older pond to Bouch, he dispatched an analysis of water from the new pond, finding that it contained 5.90 grains of solid matter and commenting "It is remarkably pure water and well fitted for boiler purposes".[21] The total distance of piping necessary for this supply was significantly shorter (just over 2 miles), and this was the supply that was chosen. The estimated cost of restoring the embankment of the pond after it had been cut through to lay the pipe was 267 pounds 7 shillings and 8 pence, and the cost of providing and laying the pipes to the station was 849 pounds 14 shillings and 4 pence.[22] Work commenced almost immediately.[23] The provision of a suitable water supply could clearly be an expensive business, but nevertheless the railway company must have saved a considerable amount of money in construction costs by using the nearer pond. The line was opened on 8 August 1861, just under a year after Richardson had submitted his analysis of the water for the new pond. It was agreed that Witham should receive 50 pounds per annum for the water.[24]

While it was obvious that neither the engineers nor anyone else employed by the early railway companies possessed the knowledge or skills to carry out water analyses, it was also the case that those in the railway industry often had no concept of the amount of work involved in analysis. This is highlighted by the example of a dispute in 1855 between the Lancaster and Carlisle Railway and the chemist Edward Frankland (see Chapter 3). The company had commissioned 13 water analyses from Frankland at a fee of 5 guineas per analysis, but on receiving the analyses, the board instructed the company secretary to write to Frankland stating that a reduced charge was appropriate in view of the number of analyses performed.[25] Frankland asked the opinion of three of his chemical colleagues (Prof. F. C. Calvert, Dr Angus Smith and Prof. Thomas Graham) before sending a blistering reply. He opened by stating that as a rule, railway directors had no knowledge of the nature of chemical analysis and hence were unaware of the amount of work involved. In this instance, he had worked for six days per week, for seven weeks. He claimed that if an engineer had spent the same amount of time surveying a branch line and had charged the same amount, the directors would have been surprised at the moderation of the charge. He drew attention to the opinion of Thomas Graham, who had served from 1841 to 1843 as the Chemical Society's first President and who had just been appointed to the important position of Master of the Mint. Graham quoted 10 guineas as the generally understood charge for a water analysis. Frankland claimed that if he had charged his own minimum daily rate of 3 guineas, his account would have been for 88 pounds rather than for 68 pounds and 5 shillings. He pointed out that performing multiple analyses resulted in a saving of neither time nor labour. He concluded by saying he thought that he had made a very moderate charge, and that when the directors were in possession of the facts they would doubtless be of the same opinion.[26] One week later Frankland received a letter from the Lancaster and Carlisle enclosing the agreed fee.[27]

Frankland (always hypersensitive in matters financial) appears to have been jealous of the fees which could be commanded by engineers. Later, in the 1870s, he would be a leader in a movement to create a professional body for chemists, based on academic and practical qualifications, with a view to securing both appropriate status and remuneration for them. Growing rivalry between chemists and engineers was one of the critical issues which led to the foundation of the Institute of Chemistry (another of the parent organisations of the Royal Society of Chemistry) in 1877, with Frankland as its first President. In the words of the first historian of the Institute, Frankland could be considered "to be practically the Founder".[28]

4.5 WATER TREATMENT

Though it became general practice for railway companies to use water analysis to decide which was the best option when selecting a water supply, sometimes (*e.g.* in limestone districts) the use of a hard water was inevitable. Indeed, a perfect water supply, containing no undesirable dissolved substances, was rarely encountered. Thomas Richardson, in his comment on an unusually good water (from Kirkby Stephen), said "It is admirably adapted for Boiler purposes, in fact I have rarely met with so pure a water and with proper management the boiler ought never to fur."[29] The key phrase here is "with proper management". All boilers had to be washed out at intervals, the point of the procedure being the removal of precipitated material before it had formed a hard deposit around the firebox and tubes. Thus a water analysis could not only aid in the selection of a water supply, but it could also give an indication of how frequently the washing out procedure should be conducted.

There were no standards or specifications for water supplies, as there were later to be specifications for materials such as firebox plate or rail steel. In general, the rule was that the softer the water, the better. By the end of the steam era, railways were softening water as a matter of routine, but in the early days only the hardest waters were candidates for treatment. Chemists were not surprisingly involved in devising methods of water softening, but in the early period the railway companies did not employ them as consultants for this purpose. The chemists appear as patentees, hoping that railways would adopt their processes.

A chemical method of softening water, involving the addition of a calculated quantity of lime water, was patented by Thomas Clark in 1841.[30] In 1843, Clark tried to persuade Brunel to adopt the process on the Great Western Railway, but was unsuccessful.[31] In 1846, William West advised against the process, pointing out that it only removed *some* of the undesirable components (the so-called temporary hardness) and that it required the installation of large settling tanks and filters. He recommended that where hard water had to be used, frequent and early "blowing through" (*i.e.* washing out) of the boiler could largely avert problems.[32]

There was, however, another possibility, which was to add substances to the water after addition to the boiler, and such techniques found some application. The most widely used method was patented in 1844 by a London medical man, Dr Louis Antoine Ritterbandt.[33] This involved the addition of the chemical muriate of ammonia (ammonium chloride) to the boiler, which dissolved any scale which had already formed, and prevented any further precipitation.[34]

The method was tested in 1845 by Ritterbandt on the London and South Western Railway under the supervision of the company's locomotive engineer, John Viret Gooch.[35] The locomotive employed, *Reindeer*, initially had a layer of scale between $1/16^{th}$ and $1/8^{th}$ of an inch in the boiler, on the tubes and on the firebox, apart from a small area which had been cleaned. During the subsequent trial, which lasted five weeks, Ritterbandt added ammonium chloride as he felt necessary, with the boiler being washed out every four days as usual. After the trial period, all the incrustation had disappeared, and the cleaned portion of the firebox was as bright as at the start. Gooch was anxious to determine whether the ammonium chloride had attacked the metal, so he drew off some water from the boiler "and had it severely tested by a practical chemist of considerable experience, but no trace of copper or iron was found". A chemist called Johnson reported that he also had tested several specimens of water from boilers in which ammonium chloride had been used, but had been unable to detect the presence of metals.[36] Although this early trial was apparently successful and the method received some application, it was found that the prolonged use of ammonium chloride caused boiler corrosion.[37] A number of other proprietary "boiler compounds" were used by some companies to add to very hard water in the boiler. These generally consisted of soda (to soften the water) boiled with a material, such as starch, treacle, sawdust, *etc.* The aim was to coat each precipitate particle at the moment of its formation with a gelatinous material which would prevent it sticking to the boiler. A serious start to the resolution of the problems associated with hard water only occurred towards the end of the nineteenth century with the development of the first automated softening plants (see Chapter 8). In the meantime, railway companies were reduced to securing the best (or least bad) water supply, the choice being made with the aid of a chemical analysis. However, once a company had obtained an analysis of a water supply (either existing or potential), there was rarely the need for it to be repeated. Thus although water analysis was of vital importance to the railways, it never generated enough work to justify a company appointing a chemist to its full-time staff.

4.6 CONCLUSION

From 1835, the railway companies were "testing the waters" in two senses. They were gauging the suitability of water supplies for their locomotives, but they were also assessing the value of the services offered by the chemists who performed the analyses. It was only a short space of time before the chemists had proved their worth, and soon they were being asked to perform other tasks. It is to these further roles of the chemical consultants in the early railway industry that we must now turn.

REFERENCES

1. W. West, *J. Roy. Inst.*, 1830–1, **1**, 38.
2. W. West, *Minutes of Proceedings of the Institution of Civil Engineers*, 1846, **5**, 182.
3. Minutes, Liverpool and Manchester Railway Board, May 1826–June 1830 and June 1830–Feb. 1833 (TNA, RAIL 371/1–2).
4. W. V. Farrar, in *The Henrys of Manchester and Other Case Studies*, ed. R. L. Hills and W. H. Brock, Variorum, Aldershot, 1997.
5. Ref. 3.
6. Obituary notices, *J. Chem. Soc.*, 1853, **5**, 158; *Minutes of Proceedings of the Institution of Civil Engineers*, 1851–2, **11**, 112. Biographical sketches: *Proceedings of the Yorkshire Geological and Polytechnic Society*, 1889, **10**, 238; *Chem. Drug.*, 1960, **174**, 71.
7. W. West, *Q. J. Sci.*, 1823, **15**, 82; *idem, ibid.*, 1827 (II), 21.
8. Letter, W. West to R. B. Dockray, 23.09.1835, Stockton and Darlington Railway letter book (TNA, RAIL 667/1068).
9. Meetings of 27.07.1839 and 24.08.1839, Hull and Selby Railway Directors' Minutes, 1838–1840 (TNA, RAIL 315/7).
10. Guard Book compiled by Daniel Gooch of miscellaneous papers on railways and scientific subjects, 1838–1851, p. 18 (TNA, RAIL 253/334).

11. Letter, E. Bury to Capt. Morsom, 18.04.1838, London and Birmingham Railway Correspondence, 1838 (TNA, RAIL 384/280).
12. Minutes 18.04.1842, Newcastle and Carlisle Railway, Minutes and Reports, 1842–6 (TNA, RAIL 509/7).
13. Ref. 2.
14. Four sites were selected for watering stations at various points on the line. Letters of Thomas Bouch to James Affleck 26.09.1859, and Thomas Bouch to A. L. Nimmo, 26.09.1859, South Durham and Lancashire Union Railway, Letters from Thomas Bouch, 1856–63 (TNA, RAIL 632/70, File 4).
15. C. D. Watkinson, Thomas Richardson, *Oxford Dictionary Natl. Biography*; M. Byrne, Thomas Richardson: his contribution to chemical education, *Durham Res. Rev.*, 1974, 7, 944; C. A. Russell, in *Dictionary of Nineteenth-Century British Scientists*, ed. B. Lightman, Thoemmes Continuum Press, Bristol, 2004.
16. Letter, Richardson and Browell to William Bouch, 02.08.1860, Railways: Scrapbooks of Early History, Letters and Cuttings, *ca*. 1813–1885, vol. 1, p. 65 (NCL, L 656.2).
17. Survey of the Lartington Estate, with plans of farms, acreages, croppings, *etc.*, 1856–1861 (DCRO, D/HH/7/4/7).
18. A list of the original directors is given in G. Brown, *Summary of Facts and Expenditure of the South Durham and Lancashire Union Railway*, Ward, Barnard Castle, 1863.
19. Letter A. L. Nimmo to G. Brown, 20.07.1860, South Durham and Lancashire Union Railway, Water Supply at Barnard Castle, 1860 (TNA, RAIL 632/45).
20. Sections and elevation of dam for proposed fish pond at Lartington, Sept 1858 (DCRO, D/HH/7/4/245-246 and 248).
21. Letter, Thomas Richardson to William Bouch, 15.08.1860; ref. 16, p. 65.
22. Letter, J. Anderson to G. Brown, 24.08.1860; ref. 19.
23. Note of cash paid to Mr Anderson on Account of Water Supply from Lartington Lake to Barnard Castle, Nov. 1860–Apr. 1861; ref. 19.
24. Draft abstract of grant of easement for water supply from Rev. Thomas Witham to the South Durham and Lancashire Union Railway, 16.03.1861 (DCRO, D/HH/7/4/494).

25. Letter, S. N. Borlam to E. Frankland, 02.06.1855 (RFA, OU mf 01.01.0265).
26. Letter, E. Frankland to S. N. Borlam, 11.06.1855 (RFA, OU mf 01.08.0266).
27. Letter, S. N. Borlam to E. Frankland, 18.06.1855 (RFA, OU mf 01.01.0264).
28. R. B. Pilcher, *The Institute of Chemistry of Great Britain and Ireland. The History of the Institute: 1877–1914*, Institute of Chemistry, 1914, p. 50.
29. Letter, Thomas Richardson to William Bouch, 24.10.1859; ref. 16, p. 61.
30. T. Clark, *Repertory of Patent Inventions*, 1841 (N.S.), **16**, 225.
31. Michael Faraday wrote to Brunel, saying that Clark wished to speak with him "regarding the fur or earthy deposit in locomotive engines", adding that he was providing Clark with a letter of introduction. Letter, M. Faraday to I. K. Brunel, 15.08.1843; Letter 1513 in F. A. J. L. James, *The Correspondence of Michael Faraday*, Institution of Electrical Engineers, London, vol. 3, 1996. No details of the meeting survive, but the Great Western Railway did not adopt Clark's process.
32. Ref. 2, pp. 190–1.
33. L. A. Ritterbandt *Br. Pat.*, 10409, 02.12.1844.
34. In terms of a modern chemical equation, ammonium chloride would act upon precipitated calcium carbonate thus: $CaCO_3 + 2NH_4Cl \rightleftharpoons CaCl_2 + (NH_4)_2CO_3$. In boiling water the ammonium carbonate would dissociate into carbon dioxide and ammonia, thus driving the equilibrium to the right. Calcium hydrogen carbonate (bicarbonate) in solution would behave similarly, yielding ammonium hydrogen carbonate which would likewise decompose to give carbon dioxide and ammonia.
35. Minute 22.08.1845, London and Southampton Railway Locomotive Way and Works Committee Minutes, 1839–1845 (TNA, RAIL 412/4).
36. The trial is described by Gooch and Ritterbandt in ref. 2, pp. 195–202. Johnson's remarks are on pp. 208–9.
37. The method and its deficiencies were commented on as late as 1890; C. L. Bloxam, *Chemistry Inorganic and Organic*, Churchill, London, 7th edn, 1890, p. 49.

CHAPTER 5

More Work for the Chemical Consultants

5.1 INTRODUCTION

Once the chemists had proved their worth in water analysis, it is not surprising that almost immediately the railway companies asked them to perform other tasks. The railways needed huge quantities of fuel, and chemists were able to analyse samples from different potential suppliers and advise on their relative qualities. The railways created an unprecedented demand for timber preservation, and chemists were called upon to advise on the merits of the various methods available and to analyse the preservatives employed. There was also a need to allay the fears of a still sceptical travelling public who were concerned that they might be adversely affected by fumes when passing through long tunnels. In the latter part of the nineteenth century, those companies which still employed consultants asked them to analyse a significantly greater range of materials.

5.2 FUEL ANALYSIS

In 1830, the year the Liverpool and Manchester Railway opened, the famous German chemist, Justus von Liebig, devised a method which enabled the percentages of the elements present in organic (*i.e.* carbon containing) compounds to be determined much more quickly than hitherto. During his sojourn in Germany, Thomas Richardson (see Chapter 4) worked in Liebig's laboratory, and one of his research projects was the application of Liebig's new method to the analysis of coal. On his return to Britain in 1838, Richardson published analyses he had performed on eight varieties of coal from various British locations. The composition of the samples was reported in terms of carbon, hydrogen, nitrogen and oxygen (reported together), and ash.[1] The method was equally applicable to coke. Richardson also described a method by which the calorific value of a coal could be computed from its elemental analysis. The following year, William West pointed out that coal usually contained sulfur, but that this was never included in the analyses. He read a paper in which he described a method for the determination of sulfur, and he gave the results he had obtained for several British coals.[2] Just as many railway companies obtained an analysis of a water supply before placing a contract, some now consulted a chemist before placing a contract for fuel.

The fuel in general use by the early railways was coke rather than coal. This arose because of objections to coal-burning locomotives on the grounds that they generated a great deal of smoke. In consequence, an early piece of legislation stipulated that railway locomotives should "consume their own smoke".[3] Good quality coke generated almost no smoke, so it was used in place of coal until around 1860, when a method of burning coal cleanly in the firebox was developed by Matthew Kirtley and Charles Markham of the Midland Railway by fitting the firebox with a brick arch, and the firedoor with a deflecting plate.[4] Many railway companies, like the Great Western Railway, coked the coal themselves. They were naturally concerned to construct their coke ovens to the best design and to buy coal which produced the best quality coke. One of Brunel's notebooks contains chemical analyses, dated 1840, of the coke produced from three varieties of coal in terms of the carbon, water and volatile matter, sulfur, and incombustible matter which they contained. It is not recorded who performed these analyses.[5]

In 1848, Daniel Gooch (locomotive superintendent of the Great Western Railway) organised a more extensive set of experiments, apparently prior to the company placing a large coal contract. Chemical analyses were performed on 11 samples of coke by

Thomas Richardson, on nine samples by Thomas Spinney of Cheltenham, and on four samples by William Herapath. All were analysed for carbon, ash, and sulfur, with Spinney reporting on the volatile matter as well. While a chemist would almost always include with a water analysis his opinion as to the suitability of the water for use in locomotives, only Herapath made a comment on these coke analyses: "The percentage of pure coke (*i.e.* carbon) will give the relative steam generating power of the samples *very nearly*".[6] It would have been understood that the quantity of ash should be as low as possible, and that sulfur was an undesirable component, on account of the corrosive and noxious nature of the gaseous products. In addition to the chemical work, further tests were carried out on some of these cokes to establish how well locomotives performed when they were used for firing, with measurements being made of how much coke was consumed and how much water was evaporated per mile using six different locomotives. Experiments were also performed to determine how much coke was yielded by the corresponding coal from four different types of oven.[7]

Thus it is clear that by the 1840s, chemical analysis, along with actual experience of the fuels in service, was being used to decide which supplier should get a particular fuel contract. Chemical analysis therefore helped a railway company to get the best value for money. As with water analysis, it made sound economic sense for a railway company to analyse potential sources of fuel, but an individual company would only require fuel analyses on an infrequent basis. Once a company had decided on a provider and placed a contract, there would be no need for further analyses unless for some reason it decided to seek another supplier. Such a situation arose at the London and North Western Railway in 1853. The company was faced with a rise in the price of the Durham coal which it purchased. Since this threatened to increase its annual bill by £10 000, it investigated the cost and quality of alternative supplies from South Wales, and chemical analysis was called upon to assist in making the decision.[8]

5.3 THE PRESERVATION OF TIMBER

When chemists were consulting for railway companies on potential water or fuel supplies, their advice depended solely on their analyses of those consumables. In the case of timber preservatives, chemists likewise performed analytical work but, in addition, they provided more general advice, and there is one example of a company commissioning research from a chemical consultant.

A major use for timber on railways was for the sleepers to support the tracks. Wooden sleepers had been used long before the railway age, on many of the early waggonways in mines.[9] The rotting of the timber was an obvious problem and an alternative, used on many of the tramroads which preceded the railways, and indeed on some of the early railways themselves (*e.g.* the Liverpool and Manchester Railway), was to support the rails on stone blocks. However, the heavy stone blocks were found to sink slowly into soft ground, with the result that the track was distorted and the gauge altered. By using transverse wooden sleepers resting on ballast, the load was spread more evenly over the ground and in consequence there was less distortion of the track. The ride was also smoother. Those railways which initially laid their track on stone blocks soon made the change to sleepers, and many used sleepers or other wooden supports from the outset.[10] Furthermore, before the age of steel, wood was used for a larger range of constructional purposes, examples being turntables and viaducts. All this timber had to be impregnated against decay.

Although the distillation product of coal tar, creosote, was to become the timber preservative of choice from the late 1850s, in the early 1830s the most frequently employed method of preservation was that patented by John Kyan in 1832. This involved soaking the timber in a solution of corrosive sublimate (mercury(II) chloride).[11] In 1837, when the construction of the Great Western Railway was proceeding apace, Brunel wrote to Michael Faraday, asking for advice on certain aspects of Kyan's process. He wanted to know how deeply the corrosive sublimate would penetrate into the wood, whether freshly treated timber might affect iron spikes and bolts inserted into it, and whether certain other chemicals might be equally effective as timber preservatives. He added that he needed to treat between 40 000 and 50 000 loads of timber.[12] Faraday was the most appropriate scientist for Brunel to approach on this issue, for in 1833 he had lectured at the Royal Institution on timber preservation and had come out in favour of Kyan's process.[13] More recently he had performed experiments for the Admiralty to assess how deeply corrosive sublimate would penetrate into various types of wood.[14] Faraday's reply to Brunel's query has not survived, but Brunel adopted Kyanizing,[15] and a few months later Faraday sent Brunel an analysis he had performed on the sediment removed from the Kyanising tanks.[16] Shortly afterwards, Brunel sent Faraday two samples of corrosive sublimate for analysis, noting that one did not dissolve completely.[17] Faraday found that this sample was contaminated with a considerable quantity of insoluble calomel (mercury(I) chloride).[18]

The chemical work that Faraday performed for Brunel was probably done on an unpaid basis. There was evidently a bond of friendship between the two, for in 1836, when Brunel was spending much of his time away from London supervising the construction of the Great Western Railway, Faraday wrote to him, "Railroads which bring other things together have separated you and me lately but I hope they will have a contrary effect bye and by [sic]".[19] At this time Faraday was attempting to relinquish his consultancy work. He later recorded that by 1836, apart from advising Trinity House, he had abandoned such external activities.[20] Although after this date Faraday still occasionally advised Brunel on scientific matters, he made it clear that he was not prepared to act as a paid consultant.

Kyanizing was employed by many railway companies in the early days. An article published in 1839 cited 18 railway companies which were using the process: among them were the London and Birmingham, the North Midland, the London and Southampton, as well as the Great Western Railways[21] In the same year the Hull and Selby Railway adopted the Kyanizing process after receiving the advice of a chemist: "The Secretary informed the Board that he had conversed with Mr Pearsall, the Chemist, relative to Kyanizing Timber, and that he gave a very decided opinion in favour of the process".[22]

Corrosive sublimate was soon challenged by other substances. One was the evil-smelling coal tar, first exploited for a while by Lord Dundonald. This was the material which condensed out when coal was heated to produce gas and coke Generally, however, much was discarded, although it was later (from 1850) exploited as a source of organic compounds. Brunel had mentioned to Faraday that he was experimenting with coal tar,[23] but a better preservative was obtained from the distillation of coal tar. It was termed "heavy oil of tar" or "creosote (oil)", although the term "creosote" was originally applied to the oil obtained by distilling wood tar. It was John Bethell who showed in 1838 that heavy oil of tar could be used as a wood preservative,[24] and in 1840, Brunel exchanged letters with him about possible trials on the Great Western and on the Bristol and Exeter Railways.[25] In 1846, Faraday and Brunel corresponded on whether compounds such as naphtha and ammonia in coal tar would be injurious if creosote and coal tar were used in combination.[26]

In the 1840s and 1850s, many other methods of timber preservation were suggested, the growth of the railway system being the spur to the invention of new treatments. Most of these probably found little or no application, but zinc chloride and copper sulfate were certainly used to some extent. However, with such a wide variety of treatments on offer, one company, the North Eastern, decided in the mid 1850s to commission a chemist to investigate the various options. The man chosen to undertake experiments, which were described as "extending over a period of some years", and to produce a report, was Thomas Richardson.[27]

Richardson listed 63 patents, granted between 1832 and 1858, which described a wide variety of both preserving agents and methods of impregnating the timber. He commented that coal tar creosote was generally acknowledged to be the most suitable material, and that Kyan's process was not now widely used on account of "the high price and the danger to which the workmen are exposed by its absorption". He described a series of experiments which he had performed at the Gateshead works of the North Eastern to compare the effectiveness of a range of preservatives. Having first modified the apparatus used for impregnating the timber, he set about treating sleepers with various agents. In these experiments he was assisted by A. F. Marreco, a prominent Newcastle chemist of Portuguese descent who subsequently became a founder member of the Institute of Chemistry and professor of chemistry at the Durham College of Science, Newcastle.

Before these experiments were complete, Richardson developed a new preservative of his own invention. Noticing piles of old sleepers lying beside the line between Newcastle and York, he suggested distilling them to make wood tar, the gas being used for lighting purposes. The wood tar, instead of being redistilled to make wood creosote, was to be dissolved in crude caustic soda, and the resulting fluid used as a preservative for new sleepers, fence posts, *etc*. Richardson was assisted by his partner E. J. J. Browell in this work, and together they patented the process.[28] Richardson suggested that the charcoal left behind in the retorts after the distillation process be employed in "steeling" the surface of rails, a technique which at that time was being employed by the North Eastern Railway.[29] The experiments to develop the process to produce the new wood preservative were also conducted at the Gateshead works, but there is no record of it being adopted by the North Eastern Railway or by any other company. No comment by any official of the North Eastern concerning Richardson's work has survived. While most of the chemical consultancy work performed for the railways was probably considered to be good value for money, it is unclear whether this project falls into that category.

Coal tar creosote eventually became the substance of choice for wood preservation, not least because the huge expansion of the gas industry in the nineteenth century increased its availability and reduced its cost. By the 1880s, the firm of Burt, Boulton and Haywood was described as being the largest tar distillers in the world, a success achieved through supplying the railway industry.[30]

5.4 AIR QUALITY IN TUNNELS

In the examples quoted so far, the work of the chemical consultants is unlikely to have come to the attention of the general public. However, there was an occasion when reports produced by chemists and medical men were published in order to reassure potential travellers who were fearful of one aspect of the new mode of transport. This was travelling through tunnels. A railway guide, published around 1840, commented that until recently the greatest fear of the public had been of being crushed by the collapse of the tunnel, but that another concern had been the possible danger of breathing air which had been polluted by the locomotive.[31] On the very early railways, locomotives were not used in tunnels, which all happened to be on gradients. Ascent was by cable haulage powered by a stationary steam engine (see Figure 5.1), and descent was by gravity under the control of a brake wagon. The situation changed in 1836, when the Leeds and Selby Railway opened with a level 700 yard tunnel through which the trains were hauled by locomotives. John Herapath (cousin of William Herapath who had analysed

Figure 5.1 The Moorish arch on the Liverpool and Manchester Railway seen from the East. The chimneys from the two stationary engines which pulled the cables up the incline in the tunnel behind are seen rising behind the turrets of the arch. (Penny Magazine, 1833).

water and coke for the Great Western Railway) had recently become editor of the *Railway Magazine*, and he used his position to campaign vigorously against the use of locomotives in tunnels on the grounds of the danger to passengers' health.

Chemical analysis was called into play in an attempt to resolve the issue. In 1837, John Davy and R. W. Rothman reported on the air in the Leeds and Selby tunnel, "Chemically examined, its composition appears to be the same as that of the atmosphere, even after repeated transits of locomotive engines." John Davy was assistant inspector of army hospitals, but had earlier performed chemical work with his brother Humphry, principally on chlorine and its compounds, discovering phosgene.[32] Rothman was a Fellow of Trinity College, Cambridge. Published with the Davy/Rothman report was another by James Williamson, senior physician at Leeds General Infirmary, who maintained that "the vapour, smoke and the gaseous results of combustion can never exist in such proportions as materially to deteriorate the air". He also expressed the opinion that excursions by train were positively beneficial for health, even to persons "labouring under the slighter forms of pulmonary irritation".[33]

Another report on the air in the same tunnel was produced in March 1837 by David Boswell Reid (1805–1863).[34] As well as being a lecturer in chemistry at Edinburgh, Reid was an expert on ventilation. He had recently acted as an advisor on the ventilation of the Houses of Parliament, and was subsequently to write a book on the subject.[35] Reid's report on the Leeds and Selby tunnel described how he had measured the concentration of carbonic acid gas (carbon dioxide) in the air, and had found it to be considerably less than 1%. He had also calculated that the amount of carbonic acid formed from the coke consumed as the locomotive passed through would be insignificant. He pointed out that the hot carbonic acid gas would float above the roofs of the coaches until it escaped. He detected no other impurity in the air of the tunnel apart from the carbonic acid.[36]

A report on the Primrose Hill tunnel, then under construction on the London and Birmingham Railway, was published at the same time. This was produced by four eminent medical authorities, and by Richard Phillips, who was lecturer in chemistry at St. Thomas's Hospital.[37] The authors described how they had been conveyed into the partially completed tunnel by a train which then remained stationary for 20 minutes before returning. They reported that the air was "apparently unaffected by steam or effluvia of any kind: neither was there any damp or cold perceptible." They concluded "...the apprehensions which have been expressed that such Tunnels are likely to prove detrimental to the health, or inconvenient to the feelings of those who may go through them, are perfectly futile and groundless".[38]

The question remains as to who commissioned these reports. No reference to them occurs in the minutes of the Leeds and Selby or the London and Birmingham Railways. However, there is no doubt that the authors were paid for their trouble, and therefore rank as consultants. Herapath republished all the reports, and added his own scathing remarks. Commenting on Reid's report, he said "We beg to whisper to the Doctor and to others, that fees sometimes go a great way to forming opinions. In France, gold taken internally, has lately been found to be very serviceable in certain diseases of the body; we now fear in England, if taken in the pocket, it may occasionally produce sad aberrations of the mind."[39]

The reports may have been commissioned by individual directors or shareholders, acting in a private capacity. Whatever their origin, they show that the opinions of chemists, alongside those of medical practitioners, were being used to reassure the public on an aspect of the new mode of transport about which they were apprehensive. Herapath promised to issue further challenges to the authors of the reports, but he did not do so. By the end of 1837, a number of much longer tunnels had been authorised, and Herapath presumably felt that the battle had been lost.

5.5 THE SWANSONG OF THE CHEMICAL CONSULTANT

Once the railway companies started to appoint their own full-time chemists (see Chapter 6), the days of the chemical consultant were numbered. In the latter part of the nineteenth century, however, there was a brief period in several companies when, before they decided to employ their own chemists on a full-time basis, their consultants were asked to report on a wider range of materials. For example, in 1870 the Newcastle chemist A.F. Marecco prepared a report for the North Eastern Railway on oils for signal lamps,[40] in 1873 the Midland Railway obtained a chemical analysis on a batch of steel rails which were breaking in service,[41] in 1883 Edward Frankland analysed for the London and South Western Railway materials used in lubricants (tallow, lard oil and rape oil) and materials used in paints (linseed oil, turpentine and pigments),[42] while in 1886 the Lancashire and Yorkshire Railway asked their consultant, Dr Charles Anthony Burghardt, to report on india rubber springs, patent compounds to inhibit boiler corrosion and various lubricating oils.[43] By the early twentieth century, however, almost all such analytical work, as well as a considerable amount of other chemical work, was being performed by company chemists in company laboratories, and as a result the independent chemical consultant had almost completely disappeared from the railway scene. Some work continued to be given to consultants by smaller companies who never appointed their own chemist, examples being provided by a steel analysis for the Furness Railway in 1899 and by a water analysis for the Metropolitan Railway in 1902.[44] However, many of the smaller companies without laboratories were allied in one way or another to larger undertakings, and were able to call upon the chemist of the larger company if necessary. An example is provided by the Midland and Great Northern Joint Railway (MGNJR) in 1903, when the chemist of the Great Northern performed several analyses of water and white lead for the engineer of the MGNJR.[45] Thus although a small amount of chemical work continued to be performed on a consultancy basis, there is no doubt that from the late 1880s onwards the railway chemical consultant was a dying breed.

5.6 CONCLUSION

In the early days of railways, the work performed by chemical consultants quickly became indispensable to railway operations. Nevertheless, in 1864 the amount of chemical work required by the railways was still insufficient for any company to consider carrying out the work itself. However, the introduction of a new technology in that year meant that, for the company concerned, a chemist became a key member of its permanent staff.

REFERENCES

1. T. Richardson, *Trans. Nat. History Soc. Northumberland*, 1838, **2**, 401.
2. The report of the meeting at which the paper was read is in *Proceedings of the Geological and Polytechnic Society of the West Riding of Yorkshire*, 1839–42, **1**, 48 (the paper itself was not published).
3. This stipulation was first made in the 1826 Act for the Liverpool and Manchester Railway, and was incorporated into the Acts for subsequent railways. E. L. Ahrons, *The British Steam Railway Locomotive 1825–1925*, Locomotive Publishing Co., London, 1927, p. 131.
4. Ref. 3, pp. 134–6.
5. I. K. Brunel, *Volume of Facts,* 1838–1844, p. 109, 13.12.1840 (TNA, RAIL 1149/10).
6. All these analyses are contained in *Guard Book* compiled by Daniel Gooch of miscellaneous papers on railways and scientific subjects, 1838–1851, pp. 36–7 (TNA, RAIL 253/334).

7. Minute 23.03.1848, GWR General Committee, Mar 1847–Jul 1848 (TNA, RAIL 250/125).
8. E. Watkin and J. E. McConnel, Report on Welsh coal and coke supply, 1853, *LNWR: A Collection of Private and Confidential Reports, Circulated to Directors, 1838–1853*, (SML, 625.1 [410]).
9. A description of the construction of a waggonway is given in a guide published in 1807 called *The Picture of Newcastle upon Tyne*. See P. J. G. Ransom, *The Archaeology of the Transport Revolution*, Guild Publishing, London, 1984, p. 31.
10. Initially the Great Western Railway used longitudinal wooden beams with intermittent cross-ties rather than transverse sleepers.
11. J. H. Kyan, *Br. Pat.* 6253, 31.01,1832.
12. Letter I. K. Brunel to M. Faraday, undated (*ca.* June 1837), no. 1006 in F. A. J. L. James, *The Correspondence of Michael Faraday*, Institution of Electrical Engineers, London, vol. 2, 1993.
13. M. Faraday, *On the Practical Prevention of Dry Rot in Timber*, lecture delivered at the Royal Institution, 22.02.1833, John Weale, London, 1836.
14. Letter M. Faraday to J. R. Clark (Admiralty official), 29.05.1835, ref. 12, no. 798.
15. GWR correspondence regarding engineering work, 1835–9 (TNA, RAIL 1008/62).
16. Letter, M. Faraday to I. K Brunel, 19.10.1837, ref. 12, no. 1044.
17. Letter, I. K. Brunel to M. Faraday, 13.11.1837, *I. K. Brunel's Letter Book*, Aug. 1837–Mar. 1838, f.132 (TNA, RAIL 1149/3).
18. Entry 02.12.1837 in RI Laboratory Notebook, 1830–1861, (RI MS, HD 8b, p. 100).
19. Letter, M. Faraday to I. K. Brunel, 11.11.1836, ref. 12, no. 949.
20. A chart drawn up by Faraday detailing the dates at which he relinquished various activities is reproduced in H. Bence Jones, *The Life and Letters of Faraday*, Longmans, London, 2nd ed., vol. 2, 1870, p. 112.
21. J. Herapath, *Railway Magazine*, 1839, **6**, 468.
22. Minute of 24.08.1840, Hull and Selby Railway Directors' Minutes, 1838–1840 (TNA, RAIL 315/7). It was Thomas Pearsall who performed the water analyses for the Hull and Selby (ch. 4). In 1831, he was described as "chemical assistant

in the laboratory of the Royal Institution". See T. J. Pearsall, *J. Roy. Inst.*, 1830–1, **1**, 77–83 and 267–81; 1831, **2**, 49–61.
23. Ref. 12.
24. S. B. Boulton, *Minutes of Proceedings of the Institution of Civil Engineers*, 1883–4, pt 4, **78**, 97.
25. Letters, I. K. Brunel to J. Bethell, 29.05.1840, 30.03.1841, 21.06.1841, ref. 17, ff.79, 224, 250.
26. Letters, M. Faraday to I. K Brunel, 17.06.1847; I. K. Brunel to M. Faraday, 18.06.1847; M. Faraday to I. K. Brunel, 18.06.1847. Letters 2000–2, in F. A. J. L. James, *The Correspondence of Michael Faraday*, Institution of Electrical Engineers, London, vol. 3, 1996.
27. Pamphlet: T. Richardson, *Report on the Preservation of Timber*, 1858 (NCL). The North Eastern presented this pamphlet to the Royal Agricultural Society by whom it was published (without the Appendix listing the patents): *J. R. Agric. Soc. Engl.*, 1859, **20**, 1.
28. J. Richardson and E. J. J. Browell, *Br. Pat.* 1036, 13.04.1857.
29. This was a process, invented by Isaac Dodds, by which the surface of a wrought iron rail was case-hardened by heating with charcoal. It was used at the time by the North Eastern, but was discontinued before 1875. Minute 7198, 08.10.1875, NER Board, 1875–80 (TNA, RAIL 527/15).
30. M. Mills, *The Early Gas Industry and its Residual Products in East London*, unpublished PhD thesis, Open University, 1995, p. 110.
31. A. Freeling, *The London and Birmingham Railway Companion*, Whittaker, London. 2nd edn, no date (1840?), pp. 33–6.
32. J. R. Partington, *A History of Chemistry*, Macmillan, London, vol. 4, 1964, p. 73.
33. Pamphlet: *Railway Tunnels. Leeds and Selby. Reports of Dr Williamson, Dr Davy and Dr Rothman*, W. Clowes, London, no date (TNA, RAIL 1015/1/11). The identical reports are dated 19 and 21 February 1837 in another version of this pamphlet also containing a report on the Primrose Hill tunnel (see ref. 38 below).
34. For Reid see R. G. W. Anderson, *The Playfair Collection*, Royal Scottish Museum, Edinburgh, 1978, p. 38.
35. D. B. Reid, *Illustration of the Theory and Practice of Ventilation*, Longman, Brown, Green and Longmans, London, 1844.

36. Pamphlet: *Report on the Atmosphere of Tunnels Founded on Chemical Analysis*, D. B. Reid, Edinburgh, 13 March 1837, (MCL).
37. For Phillips, see C. A. Russell, *Chem. Br.*, 2001, **37**, 44; obituary notice: *J. Chem. Soc.*, 1853, **5**, 155.
38. Pamphlet, *Railway Tunnels. Primrose Hill. Report of Dr. Paris, Dr. Watson, Mr. Lawrence, Mr. Phillips and Mr. Lucas. Leeds and Selby. Reports of Dr Williamson, Dr Davy and Dr Rothman*, William Clowes, London, 19 and 21.02. 1837 (MCL).
39. J. Herapath, *Railway Magazine*, 1837, **2** (New Ser.) 257.
40. Minutes 65, 68, 73, 74, 77 (1870), NER Stores Committee, 1869–1877 (TNA, RAIL 527/53).
41. Minute 182, 04.06.1873, MR Board, 1873–1876 (TNA, RAIL 491/22).
42. Minutes 59 (14.03.1883), 137 (11.04.1883), 466 (18.07.1883), 583 (16.08.1883), LSWR Engineering and Stores Committee, 1883–1884 (TNA, RAIL 411/43).
43. L&YR Committee Minutes 13.10.1885-10.02.1886 (TNA, RAIL 343/298, p. 534). Committee Minutes 16.03.1887-13.07.1887 (TNA RAIL 343/306, p. 127).
44. The Furness reported analytical figures it had obtained (from an unnamed consultant) on a broken steel rail to a Board of Trade enquiry on the strength of steel rails, Parliamentary Accounts and Papers, 1900, **30**(1), *Railways*. The analysis for the Metropolitan was carried out at "Laboratories, 14 Colville Road, London W.", and was copied on 29.12.1904 into GWR laboratory Notebook, "Waters", 17.03.1904–11.07.1905, with the figures rearranged for comparison with GWR analyses (WSRO Acc No 2515).
45. Reports dated 06.01.1903, 19.03.1903, 28.04.1903, 08.06.1903, GNR, work done by Analyst's Dept., 1903 (TNA, RAIL 236/288/11).

CHAPTER 6
A New Breed of Railwayman: The Railway Chemist

6.1 THE TESTING OF STEEL

The year 1864 was to prove momentous for one large railway, and indirectly for the others as well. Yet it is safe to say that very few people were aware of what had come upon them. This was the appointment of the first full-time railway chemist, at the LNWR works at Crewe. It came about as follows.

A most perennial problem on the infant system was that of the so-called "permanent way". Prior to the 1860s it was anything but permanent, some lengths of wrought iron rail needing replacement every few months. When that happened, it was an expensive business, involving not only new materials and workmen's time to relay them, but also closure of the line and cancellation of revenue-earning trains. A significant breakthrough emerged in 1856 at the hands of an inventor named Henry Bessemer.[1]

Realising that the presence in pig iron of considerable amounts of carbon helped to cause brittleness, Bessemer tried to burn off the carbon from the surface but with little success. He then conceived the idea that air could be blown through the crude molten metal, but was unprepared for the violence of the reaction and the

---

Early Railway Chemistry and its Legacy
By Colin A. Russell and John A. Hudson
© Colin A. Russell and John A. Hudson 2012
Published by the Royal Society of Chemistry, www.rsc.org

brilliance of the issuing flame. In little time he devised his own "converter" with holes at the base for injection of air, and a pivot to enable the subsequent decantation of the molten metal into proper receivers.[2] The initial product he obtained was really a form of wrought iron (with very little carbon). Other materials, such as small amounts of carbon and manganese, could be then be added in controlled quantities to form the kind of steel required.

Bessemer was no railwayman, nor was he a metallurgist, but he realised that "Bessemer steel" was likely to be serviceable in rails, and in addition to patenting his process he persuaded the Midland Railway authorities to lay some of his rails at a place in Derby where wrought iron had merely lasted for six or so months. The experiment was highly successful, with the track still being adequate six years later.

Now, it so happened that Bessemer had been extremely fortunate in his first experiments, for he used pig iron from Blaenavon which was very low in phosphorus. For such material his process was supremely appropriate. If the iron contained phosphorus, however, the results were disastrous. Much of the iron made in the north east, and many other places, was derived from ores rich in phosphorus (phosphates being associated with the original iron ore). However, the iron ore in areas such as west Cumbria contained no phosphorus, and so this became an early stronghold of the Bessemer steel industry.

Nevertheless, for all its limitations, Bessemer steel arrived at a most propitious time for one company, not the Midland, but the London & North Western Railway (LNWR). The rapidly expanding LNWR had experienced much financial pressure in the 1850s. One division (the southern) had replaced most of the coke by coal as a locomotive fuel as a means of economising, and was one of the first to do so. Then, in 1852, the maintenance *of the whole track* became for the first time a matter of *company* (not divisional or even area) control.[3] This was towards the end of the highly efficient management of Mark Huish (whose stringent financial reforms had even granted the firm of W. H. Smith the right to advertise and sell newspapers at his stations!).

In the early 1860s, Richard Moon,[4] a director (and since 1861 managing director) of the LNWR, sensing that his former adversary Huish was safely out of the way, started a number of moves to make the company more self-sufficient. He had already, in 1862, established the company's brickworks. Its products were used in erecting all new works buildings and a number of houses in Crewe as well.[5]

After Huish's departure in 1858, with a unified track policy in place, the time was right to test Bessemer steel rails, and this was done at Crewe in 1861 and Chalk Farm in 1862. They emerged successfully. So, in 1863 a licence was obtained from Bessemer enabling the company to make steel at its own plant (several hundred tons per week). However, the cost of steel was still double that of wrought iron, and so initially the steel was mainly used for locomotive parts. It was only the surplus steel which was used for rails, and even so the rails that were manufactured in the early years of the steel plant were steel-headed wrought iron rails.[6] At Crewe the Bessemer process continued until 1901.

The opening of the steel plant created a demand for a chemist to be regularly on duty to monitor the composition of both the pig iron entering the converters and the steel which emerged. So, in late 1864, the first full-time chemist was appointed to a railway company; the railway chemist had arrived.

6.2 CHEMISTS AT CREWE

The first railway chemist appointed to Crewe in 1864 was E. Swann, who stayed for just over two years and was succeeded by John Dods (1867–1870). Their signatures appear in analysis books, but otherwise little is known about either man, neither is there any evidence for a fixed salary for either, but they were presumably being paid by the week. Their origins are unclear, but it is likely that Swann at least had had some association with Chester College[7]. This was primarily a Diocesan training college for teachers but had recently established a "Science School" whose master had performed some water analyses for the LNWR on a consultancy basis, and doubtless transmitted some of the influence of the young chemist, William Crookes, who had taught there in 1855–6. It was Crookes who had established the teaching laboratory at Chester where the water analyses were performed.[8]

Clearly some kind of laboratory would be required at Crewe, an entirely novel feature of a railway works. On 14 July 1864, the proposal that an unoccupied cottage at the old Gas Works should be fitted up was made in Ramsbottom's report to the Locomotive

Committee.⁹ This original accommodation could hardly be described as palatial, and when three years later it was decided that it should be demolished to make way for a new saw mill, two nearby cottages were converted.¹⁰

As would be expected, analysis of pig iron and steel featured prominently in the early days. Almost as quickly the need for water analysis became apparent, hitherto a traditional job for locomotive needs performed by consultant chemists. Water intended for domestic use was also analysed; it was found in December that the supply piped to the house of the chief mechanical engineer (CME), Ramsbottom, contained too much lead! Water from one of the company's own sources was supplied to the town until a more satisfactory domestic supply was obtained from Whitmore in 1864.¹¹

In addition to his routine analytical work, Swann immediately commenced teaching at the Crewe Mechanics' Institute, an institution largely under the control of the LNWR, and founded in 1845. Almost all the chief officers at Crewe had been connected with it. Through the work of an employee at the rail mill, W. M. Moorsom¹² and his friend James Stewart, both Cambridge graduates, the Crewe Mechanics Institute was to play an important part in the University Extension Movement.¹³

Wherever Swann had learned his chemistry, he was keen to pass it on to the assembled youths. Passing through the Mechanics' Institute were several who were already employees of the works as engineering apprentices, most notably among them was Joseph Reddrop. Changing subjects, he entered the chemical laboratory, becoming assistant chemist to Swann's replacement, Dods. Upon Dod's departure for America in 1870, Reddrop succeeded him as chief chemist. Eight years later, some indication of the work that had by then accumulated and was conducted by Reddrop and others under his direction is contained in his application for Fellowship of the new Institute of Chemistry:¹⁴

> Analyses of materials required for the production of Bessemer steel,- including iron ores, fluxes, fuels, slags, sand, fire-clay, pig iron, ferro-manganese, and steel.
> Analyses of the more common metals (having special reference to the impurities) including lead, copper, tin, antimony and zinc, also their various alloys, and bronzes, *etc.*
> Analysis of water for locomotive and domestic use.

A New Breed of Railwayman: The Railway Chemist

Analysis and testing of various articles of commerce,- including soda-ash, alkalis, acids, white lead, *etc*.
Analysis of waste products from the works.

This list reflects a much wider activity of the laboratory, especially the penultimate item notably in testing some of the items carried on goods trains, serving the chemical industry that was now thriving nearby. By 1882, the staff complement of the laboratory was seven, which included a photographer and a laboratory attendant. A photograph of them has survived (Figure 6.1).

Reddrop was the first incumbent of recognised chemical stature, as was acknowledged by his transfer to the salaried list (at £150 *p.a.*) in 1876.[15] In 1895, with his assistant Hugh Ramage,[16] he was to develop the famous bismuthate process for estimation of manganese in steel (see Chapter 7). Together they constituted one of the first chemically formidable teams in the railways at that time, others examples being Phillips (Figure 6.2) and Harris (the latter also a protégé of Reddrop), and Archbutt and Deeley. Between them these partnerships may be said to have put railway chemistry

Figure 6.1 The staff of the Crewe laboratory about 1881. Joseph Reddrop is standing in the centre at the back. F.W. Harris, who was soon to move to Swindon, is on his right. (BRB (Residuary) Ltd.).

Figure 6.2 H. J. Phillips.

on the map. Ramage subsequently had a distinguished career outside the railway industry, becoming principal of Norwich Technical College.[17]

Reddrop retired on health grounds in 1899, and was succeeded by a former pupil at the Mechanics' Institute, and later a junior colleague in the laboratory, Frederick Tipler (Figure 6.6).[18] Reddrop promoted the work of the newly established Photographic Section, photographing not merely broken rails, failed machinery *etc.* but also views of the railway system, and is said to have to have been responsible for the later practice of adorning carriage interiors with pictorial photographs. Certainly the LNWR was the first to introduce this widespread practice. Tipler expanded the range of analytical activities in the laboratory, and invented a long-burning lamp for semaphore signals. He was instrumental in establishing a branch laboratory at Euston, and persuaded the general manager of the LNWR to write to all departments with the strongly worded message:

> In future the freest use shall be made of the laboratory and testing shop for every necessary purpose in connection with your department, and I shall be glad of your assurance that the matter shall have your personal attention.[19]

Figure 6.3 P. Lewis-Dale.

In 1920, Tipler was succeeded by Percy Lewis-Dale (Figure 6.3), who had once served as a clerk in the locomotive offices, was strongly attracted to chemistry and joined the chemistry laboratory as an assistant in 1910. From London University he graduated with a BSc, and in 1924 gained a PhD for his work on the by-products obtained during the manufacture of the oil gas used in lighting railway carriages. His subsequent career we shall hear more of later.

6.3 CHEMISTS AFTER CREWE

For twelve years, Crewe remained the solitary case of a railway chemistry laboratory anywhere in Britain, but during that time dramatic changes had taken place. Steel production had soared, and its price had come down, so that solid steel rails now became the norm. The sensational growth of the steel industry was made possible by the ability of the railways to transport the heavy materials to and from the works, and the railways themselves were the principal customers for the steel produced. Meanwhile the chemical community was becoming more conscious of its importance,[20] and clubbed together into the Institute of Chemistry in

1877.²¹ As a result, the value of the chemical community and of the individuals within it was being much more recognised. No longer could a "chemical" establishment be lightly established by a railway company.

Eventually the next step was taken by the railway company from another area of intense chemical growth: the north east of England. Here innumerable chemical works had been established, and a thirst developed for applying chemistry to other areas of industry. In addition to the famous Literary and Philosophical Society of Newcastle (founded in 1793 and still going strong), there were two additional societies devoted entirely to chemistry.²² Now one of the leading chemical industrialists was to advocate a railway laboratory in the north east, based upon the experiences at Crewe. His name was Isaac Lowthian Bell,²³ an owner of a large iron-making works at Port Clarence on Teesside; (Figure 6.4). He was also a director of the North Eastern Railway (NER). Bell was fortunate in having access to huge deposits of iron-ore as well as coal, and had done good business making wrought iron rails. However, the Bessemer process, as practised at Crewe, was unsuitable for the local ores for one simple reason: they nearly all contained phosphorus. The rapid adoption of Bessemer steel rails therefore

Figure 6.4 Isaac Lowthian Bell (1816–1904). (BRB (Residuary) Ltd).

presented a serious threat to Bell. However, wrought iron produced by the traditional puddling process occasionally had a carbon content, and hence mechanical properties, similar to that of mild steel. In 1876, the Board of the North Eastern Railway agreed to sponsor research at Bell's Port Clarence works:-

> The object of the N.E. Board encouraged by their own experience of the occasional excellence of iron as a material for their permanent way, was to ascertain whether rails made from Cleveland iron could not be obtained with regularity to equal the wear of those of Bessemer steel.[24]

Although Bell had a laboratory with a chemist in post at his ironworks, he also advocated that the North Eastern should establish its own laboratory. One was therefore fitted up, not at Newcastle but at the great locomotive works at York, the well-recognised centre of activity of the North Eastern Railway. Shortly afterwards it was moved to a small building in Gateshead. By 1900 it had moved to more suitable accommodation nearby in the Greensfield Locomotive Works, having a large general laboratory, a water room, a balance room and a separate office. It stayed there until its final removal in 1912, to rooms on the second floor of a group of offices in Darlington.

At York the North Eastern appointed a chemist who, unlike the young recruits at Crewe, was actually 41 and held an honours BSc in chemistry. This unusual choice was named Robert Routledge,[25] and his appointment probably reflected the influence of Lothian Bell whose sister Mary had married George Routledge, founder of the publishing firm and almost certainly a relation. Robert Routledge found time to publish a number of books on science[26] and doubtless at other times continued his round of chemical analysis of iron and other products at the North Eastern laboratory until his retirement in 1896. He was succeeded by his assistant, E. W. Rowley, graduate of the Durham College of Science, and eventually chief chemist until Grouping and beyond.[27]

Shortly to follow the examples of the LNWR and NER was the Midland Railway, centred at Derby. The first appointee was James Day, whose experience there and in previous posts gained him a Fellowship of the Institute of Chemistry (FIC). His short period of two years was followed by an even shorter one of 18 months, when

Joseph Bacon was transferred from the accountant's department. His early death at 40 was followed by a company advertisement for an analytical chemist, at £200 *p.a.* From the many applications received, Leonard Archbutt[28] was selected, a pupil of the well-known analyst A. H. Allen of Sheffield. The value of Archbutt's analytical work at Derby was to bring him the reputation of Britain's leading railway chemist. His analyses included alloys, oils and other related materials. He was active in the Society of Chemical Industry, the Institute of Metals, as well as the Society of Public Analysts, becoming its president in 1912.

Probably Archbutt's most lasting achievement was accomplished jointly with Richard Deeley (famous for his designs after appointment as locomotive superintendent for the notable 4-4-0 compound engines). Somewhat of a polymath, publishing several papers in the chemical literature,[29] Deeley collaborated with Archbutt in a new process for softening water (see Chapter 8) and in the textbook *Lubrication and Lubricants*[30] which was to become a standard work for years to come.

In the last quarter of the 19th century there was, if not a rush, then a movement of seven other companies to follow the example of the LNWR, the NER and the Midland Railway (MR), and appoint a company chemist. Most famous of all, perhaps, is the case of the Great Western Railway (GWR).[31] In 1882, a chemical laboratory was set up in a redundant school room in Bristol Street, Swindon. It had just become available with the inauguration of the Swindon School Board and the consequent establishment of a new school elsewhere in the town. The moving spirit was the works manager and later locomotive superintendent, William Dean. Previously, chemical work for the GWR was conducted at home by his assistant, W. H. Stanier, father of Sir William A. Stanier, the London, Midland and Scottish Railway (LMS) engineer immortalised by his Pacifics, Black 5s and other locomotives almost to the end of the steam age. The father had acquired various science qualifications by part-time study and had been encouraged by William Dean (whose confidential clerk he was) to conduct experiments at home and then in the new laboratory.[32] It was an impressive affair, having a floor area of 5570 square feet (see Figure 6.5), as compared with the 1875 square feet at Crewe. This was reported by Tipler (Figure 6.6) after a fact-finding mission in 1912.[33]

A New Breed of Railwayman: The Railway Chemist

Figure 6.5 The Swindon laboratory in 1883 soon after its opening. This is the earliest known photograph of a railway chemistry laboratory. (Bleasdale Collection, Science and Society Picture Library, Science Museum).

Figure 6.6 F. C. Tipler.

One of the very few popular accounts of a railway chemical laboratory was given by Acworth in 1890. Referring to the Swindon laboratory he said:

> Let us here notice the chemical laboratory, which is an equally indispensable part of a great locomotive establishment. Of all the different materials that are analysed here, the two most important are doubtless steel and oils. I have said something already of the importance of soft water for engines, and water undoubtedly takes a high place on the list. Every one knows that hard water fills the domestic kettle with a deposit of carbonate of lime, which the water as long as it remains unboiled is able to keep in solution. But there is another salt which troubles locomotive boilers, from which our kettles are free, and that is sulphate of lime, which is soluble in water boiling in the open air, but is precipitated by water boiling at high pressure. As for steel, it is a matter of common knowledge that a trifling percentage of phosphorus, or a few grains too much of carbon, means a brittle, and may at any moment mean a broken, crank axle, with all its possible consequences.
>
> Oils also have to be analysed, and this for two purposes: to test their viscosity whether, that is, they are too fluid, and will run through the cups and leave the motion to heat before an engine has gone twenty miles, or whether, on the other hand, they are too thick, and will refuse to run at all; and also, in the case of mineral oils at least, to see that they do not give off inflammable vapours at a temperature so low as to be practically dangerous. Here is an instance of the importance of a scientific study of this matter. Rape oil is very good as a rule for lubricating purposes. For some parts of an engine – such as for instance, the interior of the cylinders and steam-chest – it is not only unsuitable, but actually injurious. At the high temperature there prevailing, rape oil is decomposed – in plain English, turns rancid – and eating into the iron, forms a sort of oxide of iron soap. Not only is this in the long run bad for the cylinders, but the soap makes a lather which blocks the ports and prevents the free passage of the steam to and fro.[34]

In the same year as the GWR, the Caledonian Railway Company at Glasgow St Rollox set up a laboratory, bringing the total number of laboratories up to five. Before the end of the 19th century these were joined by the Great Northern, the Lancashire and Yorkshire, the MSLR/Great Central, the Great Eastern and the North British Railways. Of these the Great Eastern laboratory at Stratford, E. London, was to become the most famous. It was established by H. J. Phillips (Figure 6.2), and carried on by J. H. B. Jenkins (Figure 6.7), both of whom commenced their careers at the GWR laboratory. A few lines, mainly in the far south of England, did not appoint railway chemists until early in the 20th century (see Table 6.1; ref. 35).

By 1879, the annual production of Bessemer steel for rails in Britain exceeded 0.5 million tons, with some steelworks concentrating entirely on the production of rails.[36] By this time, Thomas and Gilchrist had discovered that a Bessemer converter if lined with dolomite (a basic double carbonate) could process pig iron made from phosphatic iron ore and would yield steel remarkably free from phosphorus. Rails were also being made by the open-hearth process. Principal centres of steel production were the North East, Sheffield, South Wales, Lancashire, Cumberland and Staffordshire. Railways in the south of England and in East Anglia had no steelworks in the areas in which they served, and until 1909 one company, the Great Eastern Railway, bought Bessemer steel rails made at Workington, and shipped them by sea to Lowestoft Harbour.[37] There was thus considerable scope for steel analyses on the railways for many years to come. And by 1900 the subjects for analysis had vastly increased.

The late entry of some of the southern lines is interesting. Partly this must have reflected a certain independence of the movements much further north. Also, freight traffic was less prominent, the distances were shorter, and above all the shadows of electrification caused much of their analytical requirements to be different.[38] The latter in particular (with questions such as should it be third-rail or overhead, which voltage, *etc*?) destroyed such unity as there was between steam-locomotive operators all round the country. In a sense, physics was beginning to join chemistry as the favoured science, and chemical analysis may have seemed slightly less urgent. However they got there in the end!

All these laboratories started on a small scale, usually with just one chemist. In most cases the accommodation soon proved to be inadequate, as the workload and the number of chemists employed increased. By the time of Grouping in 1923, the largest laboratories

Table 6.1 Railway laboratories and their chief chemists prior to Grouping.

Company	Date	Location of Lab	Chief Chemists
London and North Western	1864	Crewe	E. Swann (1864–1866)
			J. Dodds (1868(?)–1870)
			J. Reddrop (1870–1899)
			F. C. Tipler (1899–1920)
			P. Lewis-Dale (1920–1922+)
	1908	Wolverton	H. Mennell (1908–1917)
			F. Fancutt (1920–1922+)
North Eastern	1876	York	R. Routledge (1876–1896)
	1878	Gateshead	E. W. Rowley (1896–1922+)
	1912	Darlington	
Midland	1877	Derby	Day (1877–8)
			J. Bacon (1879–1880)
			L. Archbutt (1881–1922+)
Great Western	1882	Swindon	F. W. Harris (1882–1900)
			W. R. Bird (1900–1922+)
Caledonian	1882	Glasgow St. Rollox	S. Stewart (1882–1891)
			D. B. Fraser (1891–1921)
			W. P. Henderson (1921–1922+)
Great Northern	1886	Doncaster	Macfarlane (Feb–Nov 1886)
			J. W. Young (1887–1898)
			W. G. Young (1898–1922+)
Lancashire and Yorkshire	1887	Horwich	C. J. P. Fuller (1887–1918)
			H. Smith (1918–1922+)
MSLR/Great Central	1888	Manchester Gorton	H. Gripper (1888–1922+)
Great Eastern	1890	Stratford, E. London	H. J. Phillips (1890–6)
			J. H. B. Jenkins (1896–1922+)
North British	1894	Glasgow Cowlairs	Somerville (1894–1900)
			J. Jarvie (1900–1922+)
Great North of Scotland	1903 (Part-time)	Inverurie	?
			G. W. Urquhart (1914–1922+)
London and South Western	1903	Wimbledon	E. A. Dancaster (1903–1922+)
	1914	Nine Elms	Derrington (1914–1922+)
London Brighton and South Coast	1912	Brighton	F. P. Matthewman (1912–1922+)
South East and Chatham	1915	Ashford	H. Hall (1915–1922+)

Figure 6.7 J. H. B. Jenkins. (BRB (Residuary) Ltd.).

(at Crewe, Derby and Swindon) employed around 14 chemists, while the North Eastern employed six and the Great Central three. As in the case of the North Eastern, many laboratories moved several times into progressively larger premises, the exception being the Great Western, where the former school hall and adjacent buildings proved adequate up until the time of privatisation in 1996. Although several laboratories had to endure periods of overcrowding, in general the railway companies provided their chemists with excellent facilities.

It is clear that many railways were unsure how to position their new laboratory within the organisational structure of the company. Some, such as the Great Western and the London Brighton and South Coast, followed the lead of the London and North Western and placed the laboratory in the locomotive department. Others, such as the Caledonian and North British, placed them in the stores department. In some companies, such as the Great Northern and the Manchester, Sheffield and Lincolnshire, the laboratory was instituted in the stores department and subsequently transferred to the locomotive department. No records of discussions relating to the location of the chemical laboratory within a company's organisational structure have been found. The only observation that can be made is that those laboratories which remained within stores departments seem to have concentrated solely on quality control analyses and other relatively routine work. All the laboratories which came to be involved in major research projects were located in locomotive departments (see Chapter 8).

6.4 THE RAILWAY CHEMISTS: THEIR PUBLIC IMAGE

As late as 1954, ninety years after the first railway chemist was appointed, F. Fancutt, an assistant director of the British Rail Research Department, had this admission to make:

> To the general public, it is commonly surprising to learn that chemists are employed by railway organisations.[39]

But for British society as a whole the ignorance of railway chemistry is truly breath-taking. Certainly there were attempts to spell out before a general railway audience what a railway chemist did, but they were few and far between. Acworth had referred to the Swindon laboratory in 1890, and in 1898 *The Railway Magazine* carried a lengthier article by a chemist at Swindon. The author, G. E. Brown, gave a readable account of the work done and its relevance to railways, beginning as follows:

> Most people are surprised to find that practically all our great railway companies have a well-equipped laboratory for analytical chemistry, and retain the services of a staff of analysts. It is a common remark of visitors to a railway laboratory: "What does a railway company want with chemistry?" and at first sight there seems but little connection between the silence of a chemical laboratory, with its array of shining glass vessels and delicate balances, and the roar of the express as it thunders through the night with its living freight.[40]

This was the last major attempt to communicate the importance of railway chemistry that has been found, though perfunctory allusions to laboratories occasionally make their appearance. Often, however, such a facility is passed over in silence. An example is *An illustrated History of British Railways' Workshops*, purportedly going from 1825 to the date of publication (1998). This is a beautifully produced book, with many photographs, aerial views of works, and several maps. Despite these and an erudite text by a former deputy general manager of British Rail's workshop division (aged 92), railway chemists and their laboratories might never have existed.[41] This is a typical piece of railway writing which caters for enthusiasts of locomotive and carriage design and performance and many related matters but regards the necessary testing by chemists as of little interest or importance.

Why should this be so? Several explanations present themselves. They are basically these. First of all, as we have already seen, chemistry had undergone a social revolution in the 19th century, culminating in its professionalization in 1877. Hence the establishment of the first few railway chemists came at the height of the furore. Chemists were struggling to find their proper place in society, changing from a largely working-class group of part-time aspirants for chemical knowledge to the status of a new scientific profession (the first in the world). For them, railways were a relatively unimportant area of work. One consequence of professionalization is a growing gap between the professionals and the rest of the world. Later on, however, they could face the world with greater confidence, so by the 1950s Fancutt could also say:

> Members of the Royal Institute of Chemistry will already be aware of the contributions of the profession to engineering in general.[42]

Related to this is the general public's perception of chemistry itself, which means that chemistry is often passed by unnoticed. It was often true that, for one reason or another, analysis was regarded as boring, routine and a necessary evil. Chemistry is more than simply analysis, however, and later developments in Britain make it abundantly clear that railway chemistry is also about *research*, new materials, the effects of impurities, *etc*. Nevertheless, this has always been a matter of comparative indifference to those writing about railways for the general public.

Possibly the crucial element in the huge mass of public propaganda from the railways for 100 years or so is *competition*. Before 1923 individual lines were often competing with one another, and the famous Races to the North were a classic example. Although such competition was deliberately minimised by the Grouping, there were well-known cases of new companies trying to show that their service was better or faster than that of a rival. In such a situation, the things that are most prominent are subjects like speed, safety and design. And so public relations personnel made no mention of the chemical analyses of metals, *etc.* That is probably the strongest reason why the railway literature has overlooked the chemists. As we shall see in the following chapters, the many applications of modern chemistry to problems of the railways did not deserve to be ignored in this way.

REFERENCES

1. *Oxford Dictionary of National Biography*.
2. Sir Henry Bessemer, *Autobiography*, Offices of Engineering, London, 1905.
3. T. R. Gourvish, *Mark Huish and the London & North Western Railway: A Study of Management*, Leicester University Press, Leicester, 1972, p. 241.
4. *Dictionary of Business Biography*.
5. R. E. Jones, *Brief History of Crewe Laboratory*, unpublished typescript, 1961 (NRM/DA, 69).
6. B. Reed, *Crewe Locomotive Works and its Men*, David and Charles, Newton Abbot, 1982, p. 69.
7. Ref. 5.
8. The plans for the laboratory, dated 1855, have survived. (CCA/SC1/42).
9. Minute 202, 14.07.1864, LNWR Locomotive and Engineering Committee, 1864–6 (TNA, RAIL 410/241).
10. Minute 373, 10.05.1867, LNWR Locomotive and Engineering Committee, 1866–8 (TNA, RAIL 410/242).
11. Ref. 5.
12. W. M. Moorsom was the son of a LNWR chairman from 1860, Admiral C. R. Moorsom.
13. Ref. 5, pp. 3–4.
14. Reddrop's letter of application to the Institute of Chemistry, 19.01.1878 (NRM/DA, 154).
15. Ref. 5.
16. Ref. 5.
17. Obituary notices: *Proc. Inst. Chem.*, 1938, **62**, 285; *J. Chem. Soc.*, 1938 (II), 1135–6. Also J. A. Venn, *Alumni*

Cantabrigienses, Part II, CUP, Cambridge, 1953, and Alan Metters, *The Tech 1891–1991 – One Hundred years of Technical Education in Norwich*, City College Norwich, 1991.
18. Obituary, *J. Soc. Chem. Ind.*, London, 1920, **39**, 350R.
19. Ref. 5.
20. See, *e.g.*, *Chem. News* for 1876 (most issues).
21. C. A. Russell, N. G. Coley and G. K. Roberts, *Chemists by Profession*, Open University Press and Royal Institute of Chemistry, London, 1977.
22. See *e.g.* C. A. Russell, in *Lit. &. Phil. Bicentenary Lectures, 1993*, Literary and Philosophical Society of Newcastle upon Tyne, 1993, pp. 15–123.
23. *Oxford Dictionary of National Biography*.
24. Minute 7356, 13.04.1876, NER Board of Directors, 1875–1880 (TNA, RAIL 527/15).
25. Historical Register of London University; Chemical Society membership form (1872); Chemical Society list of members (1875).
26. R. Routledge, *Discoveries and Inventions of the Nineteenth Century*, Routledge, London, 1st edn, 1876; 14th edn, 1903; 15th edn, n.d.; *idem, A Popular History of Science*, Routledge, London, 1st edn, 1881, 3rd edn, 1894; *idem, Science in Sport made Philosophy in Earnest*, Routledge, London, 3rd edn, 1893.
27. Obituary, *Proc. Inst. Chem.*, 1932, 219.
28. Obituary, *Proc. Inst. Chem.*, 1935, 272; *Analyst (Cambridge UK)*, 1935, **60**, 579.
29. Two examples are R. M. Deeley, *J. Chem. Soc.*, 1893, **63**, 852; *idem, Proc. Chem. Soc., London*, 1895, **11**, 10.
30. L. Archbutt and R. M. Deeley, *Lubrication and Lubricants*, Griffin, London, 1st edn, 1900; 5th edn, 1927.
31. H. Holcroft, *An Outline of Great Western Locomotive Practice, 1837–1947*, Ian Allan, London, 3rd edn, 1971, pp. 53–4.
32. *Ibid*.
33. F. C. Tipler, Report to chief mechanical engineer (CME), cited in ref. 5, p. 7.
34. J. Acworth, *The Railways of England*, John Murray, London, 1890, pp. 275–276 (ch. 5, *The Great Western Railway*).
35. References to laboratories and chief chemists: LNWR: list of Crewe staff: R. E. Jones, ref. 5. Obituary of Mennell: *J. R. Inst. Chem.*, 1950, **74**, 411. Reopening of Wolverton lab. under

Fancutt: letter F. G. Dunkley to R. E. Jones, 12.10.1962, (NRM/DA, 351); NER: laboratory at York: minute 493, 12 04 1876, Stores Committee (TNA, RAIL 527/53). Move to Gateshead: minutes 18149, 11.10.1877 and 19290, 21.11.1878, Locomotive and Stores Committee (TNA, RAIL 527/33 and 34). Move to Darlington: minutes 6394, 11.01.1912, and 6435, 11.07.1912, Locomotive and Stores Committee (TNA, RAIL 527/59). Obituary of Rowley: *Proc. Inst. Chem.*, 1932, 219; MR: first reference to Day: minute 4380, 19.02.1877, Stores Committee (TNA, RAIL 491/204). Day's appointment terminated: *ibid*, minute 4819, 02.09.1878. Bacon appointed: minute 9705, 18.02.1879, Locomotive Committee (TNA, RAIL 491/176). Bacon's death: *ibid.*, minute 10294, 14.09.1880. Archbutt appointed: *ibid.*, minute 10357, 14.12.1880 (TNA, RAIL 491/177); GWR: staff list of Swindon laboratory, 1882–1970 (NRM/DA, 148); letter W. H. Roberts to R. E. Jones, 25.09.1962 (NRM/DA, 351); CR: obituary of Stewart: *Proc. Inst. Chem.*, 1932, 274. For Fraser and Henderson: letter, R. J. Ward to M. T. Hall, 08.02.1971 (NRM/DA, 148); GNR: first reference to company analyst: minute of 02.02.1886, Locomotive Committee (TNA, RAIL 236/198). Death of Macfarlane: minute of 18.11.1886, Executive Committee (TNA, RAIL 236/110). Appointment of J. W. Young and W. G.Young: Locomotive Department salaried staff register, 1862–1943 (TNA, RAIL 236/729); L&YR: first reference to company analyst: committee minutes, 28.10.1887 (TNA, RAIL 343/308, p. 296). Staff list of Horwich laboratory (NRM/DA, 351). Obituary of Smith; *Proc. Inst. Chem.*, 1929, 51; GCR: appointment of Gripper: Register of principal staff, 1862–1912 (TNA, RAIL 463/224); GER: date of Phillips leaving GWR for GER: staff list of Swindon laboratory, 1882–1970 (NRM, DA 148). Date of Jenkins's appointment as chief chemist at GER: letter, J. I. Hill to T. H. Turner, 25.10.1932 (NRM/DA, 601). Obituary of Jenkins: *Proc. Inst. Chem.*, 1927–8, 332 (date given for appointment as chief chemist is incorrect); NBR: appointment of chemist: minute of 15.02.1894, Locomotive and Stores Committee (NAS, BR/NBR/1/256); resignation of Somerville and appointment of Jarvie: minute of 22.11.1900, *ibid.*, (NAS BR/NBR/1/257); GNS: letter J. Jarvie to T. H. Turner, 17.04.1933, (NRM/

DA, 601); LSWR: letters A. J. S. Leiper to R. E. Jones, 04.10.1962, and R. L. Overin to R. E. Jones, 20.08.1962 (NRM/DA, 351); LB&SCR: minute 12, 20.12.1911, Board, 1911–1922 p. 38, (TNA, RAIL 414/84); SECR: letter R. L. Overin to R. E. Jones, 20.08.1962 (NRM, DA 351). Hall leaving Derby for Ashford: MR Locomotive Dept salary book, 1907–1922 (TNA, RAIL 491/1071).
36. K. C. Barraclough, *Steelmaking: 1850–1900*, The Institute of Metals, London, 1990, p. 85.
37. C. J. Allen, *Two Million Miles of Train Travel*, Ian Allan, London, 1965, p. 27.
38. For these and other related matters, see H. P. White, *A Regional History of the Railways of Great Britain, Vol. 2: Southern England*, David and Charles, Newton Abbot, 1982, especially ch. 1.
39. F. Fancutt, *J. R. Inst. Chem.*, 1954, **78**, 514.
40. G. E. Brown, *Railway Magazine*, 1898, **2**, 58.
41. E. Larkin, *An Illustrated History of British Railways' Workshops*, Heathfield Railway Publications, Royston, 1998 (2007 reprinting).
42. Ref. 39.

CHAPTER 7

The Railway Chemists as Materials Testers

7.1 INTRODUCTION – TESTING

From the earliest days of railways in Britain there was much testing of various kinds which increased with the years. At a human level, technical staff with aspirations were subjected to long periods of work experience before they were judged fit for promotion. Later, systems of apprenticeship were introduced, but anything like formal examinations as we understand them had to wait until the Mechanics' Institutes provided them. Users of the trains were sometimes catechised severely about the goods they had requested to transport, while passengers were required to behave correctly, and many were the unwritten tests they were required to pass. Unticketed or unruly passengers were ejected at the next station, and the rest were required to behave properly. You could not lean out of a window of a moving train, or use the toilet of one that had stopped. The reasons were obvious. Above your head there often ran a continuous cord in later days whose purpose was to halt the train in an emergency once it was pulled; or as one wag put it:

> If five pounds you can afford,
> Try your strength and pull the cord.

Early Railway Chemistry and its Legacy
By Colin A. Russell and John A. Hudson
© Colin A. Russell and John A. Hudson 2012
Published by the Royal Society of Chemistry, www.rsc.org

What was being tested, however, was rather more than your strength. The companies were obliged to provide a method of communication that the Board of Trade approved, but in fact they approved none. As Jack Simmons observed, "it is surprising that neither *Punch* nor the young W. S. Gilbert spotted this one".[1]

Testing of machinery since the Rainhill Trials for locomotives continued in complexity to the elaborate engine-testing machines at the large locomotive depots. Indeed, there are still people alive today who can recollect the ancient art of the wheel-tapper who came round testing the wheels of coaches by listening to the ringing note emitted when they were struck, quite different if they were cracked or damaged in some way. More elaborate mechanical tests, such as subjecting rails to severe impacts, were carried out in the test houses in the railway works.

Behind all this mechanical testing of equipment, however, lay an even more fundamental examination that could be performed only by chemists. So it causes no surprise to hear the words in 1914 of Leonard Archbutt, chief chemist of the Midland Railway, when he gave a speech to the Society of Public Analysts and Other Analytical Chemists. Archbutt was retiring after serving a two-year term as president of the Society, and he clearly felt that his fellow analytical chemists needed enlightening about the work of a railway chemist. Part of his address reads as follows:

> A chemist to a railway has very wide duties to perform. Foremost amongst them is the testing of the various materials of construction and stores. ...The quantities of materials used by the railway are so large that practically everything is purchased under contract, and the basis of the contract is either a specification or a standard pattern or sample. The chemist's first duties are, therefore, to draw up, or assist the engineer in drawing up, the specification, and to see that the standard pattern or sample is of a suitable quality; his next duty is to test the goods as they are delivered by the contractor, and see that they comply with the specification or are equal in every essential respect to the sample or pattern...The articles covered by the term "materials of construction and stores" are very numerous, and include various descriptions of iron, steel, and ferrous alloys, copper and other non-ferrous metals and alloys, used in the construction of the permanent

way, locomotives and rolling stock; creosote for the preservation of timber; coal, liquid and gaseous fuels; various kinds of lubricants and illuminants; paints and colours, turpentine, soap and alkalis, rubber goods, water for steam raising and for drinking, articles of food and drink sold by the refreshment department, and many other articles of lesser importance.[2]

These duties meant that the railway chemists had to be expert analysts. This chapter will provide examples of their materials testing work, and will demonstrate the importance of this activity to their employers. It will also describe some of the contributions the railway chemists made to analytical chemistry.

7.2 SPECIFICATIONS AND STANDARDS

As Archbutt made clear, the chemist was involved in determining the specifications for the very wide range of materials purchased by a railway company, and each company had its own set of specifications. Thus in a report of 1903, the Great Northern Railway's chemist, W. G. Young, when describing whether or not materials were satisfactory, used terms such as "our specification", "contract stipulation", and "GNR standard" for engine oils, sulfuric acid, turpentine, oil for oil gas production, zinc, white paint, *etc*.[3] The specifications would have stipulated acceptable ranges of composition. For materials where the specification was not critical to its performance, different companies often had widely differing specifications, probably based to a considerable extent on tradition.[4]

Before 1923, the only material for which a national standard was formulated was rail steel. Prior to 1900, there had been much discussion in the literature on the most appropriate composition, with the result that agreement was now beginning to emerge.[5] In 1900, in an appendix to the report of a Board of Trade Committee on steel rails, the specifications set by 11 companies were quoted, with the comment from the report's authors that (with one exception) the specifications were "such as we should have anticipated".[6] National standards for the composition of steel for rails had their origin in a committee created in 1901 jointly by the Institutions of Civil Engineers, Mechanical Engineers, Naval Architects and the Iron and Steel Institute, to standardise iron and steel sections for rails and for bridges and ships.[7] An early success of the committee was to cut the number of different patterns of tram rail section in production from 75 to five. Soon the committee turned its attention to matters other than sections, and from 1909 it was specifying the chemical composition of rail steel. This British Standard Specification, which was adopted by most railway companies, gave acceptable ranges for carbon and manganese, and maximum percentages of silicon, phosphorus and sulfur, but it did not have the force of law.

For other materials, there were no nationally agreed standards, and by gradually modifying specifications in the light of experience, the railway chemists contributed to the on-going technical development of the railways. Whilst the details of a contract between a railway company and a supplier would remain confidential, the specifications themselves were not usually regarded as being secret. Thus Archbutt, one of the leading authorities on lubrication and lubricants both inside and outside the railway industry, gave details of lubricants suitable for different applications.[8] Tipler, chief chemist of the London and North Western Railway, collected and published the specifications for various types of grease from 12 railway companies.[9] However, companies did not publish all their specifications as a matter of course.

7.3 ANALYTICAL METHODS IN GENERAL

We are fortunate in that we have a comprehensive compendium of the analytical methods in use in the railway laboratories in the last decade of the nineteenth century. This is a book, entitled *Engineering Chemistry*, written by H. J. Phillips and first published in 1891, with the third and final edition appearing in 1901.[10] Prior to 1890 Phillips had been second in command to Harris at the Swindon laboratory before moving to set up the laboratory of the Great Eastern Railway at Stratford. The preface opens with the words:

> In the Author's capacity as an analytical chemist, engaged in the laboratories of railway companies, he has frequently heard it remarked by pupils, whether in chemistry or engineering, as well as by practising engineers and others, that a work which should give precise methods for analysing and valuing the

most important materials in general use by engineers, would be of great service...Many of the methods of analysis given in the following pages, as far as strength and volumes of solutions are concerned, were worked out by the writer when acting as chief assistant in the laboratory of the Great Western Railway, under the direction of Mr. F. W. Harris, F.I.C., to whom he desires to acknowledge his indebtedness for much of his knowledge of chemical analysis. He also has to express his obligations to the authors of the various published works, papers in Proceedings of Societies, *etc.*, which have been consulted by him...

As Phillips indicated, the book discussed the analysis of just about anything that might be encountered by a chemist working in a railway company or in another engineering environment. Chapters dealt with a wide range of metals and alloys, ores, boiler encrustations, clays and slags, solid, liquid and gaseous fuels, water, oils, materials used in grease making, gasworks products, disinfectants, and explosives. All the methods described are, of course, what we would today call "classical", and indeed many were already well established in the 1890s. To quote one example, sulfur was estimated gravimetrically by oxidising to sulfate, precipitating as the barium salt, drying and weighing. The detailed procedure varied, however, depending on whether sulfur was being determined in steel, copper or coal, as each of these materials presented its own problems associated with the various substances present in the sample matrix. When Phillips refers in his preface to "working out" the methods with Harris, he meant the adaptation of some well-known procedures to specific situations. Harris and Phillips were not the only railway chemists who made contributions to analytical chemistry, for a considerable number of publications emanated from the railway laboratories, some examples of which will be quoted in the following sections.

Apart from describing analytical methods, Phillips's book drew attention to an important development pioneered in 1877 by Joseph Reddrop at the Crewe laboratory. Reddrop introduced a new system for the preparation of solutions in frequent use in a chemistry laboratory (bench reagents). At that time, chemists used equivalent weight[11] as the unit of quantity, and Reddrop made up his solutions in terms of whole number or fractional numbers of equivalent weights per litre, designating the concentrations 4E or E/2, *etc.* Reddrop did not publish his system until 1890,[12] apparently spurred on by the appearance of a paper by a German chemist advocating a similar system.[13] Reddrop's method was soon widely adopted, except the "E" designation was replaced by "N" (normal). Today the related but different "M" (molarity) system has largely replaced the "N" system, but this descendant of the measure first introduced by Reddrop at the Crewe laboratory is now in universal use.

7.4 METALS

7.4.1 Steel

It was the necessity to analyse batches of steel which led to the first chemist being employed by a railway company on a permanent basis (see Chapter 6). Although only one other company, apart from the London and North Western Railway, was to manufacture its own steel, the railways were eventually purchasing steel in such large quantities that its analysis became an important task in a railway laboratory. The best composition for steel for a given application was discovered empirically. A very mild steel (which replaced wrought iron) might only contain 0.15% carbon, whereas a very hard tool steel could contain 1.5%, and steel for other purposes had an intermediate carbon content. Most steel also contained manganese. Silicon, sulfur, and phosphorus were deleterious and kept to as low a percentage as possible. The standard specification for rail steel which was adopted in 1909 was: 0.35 to 0.50% carbon, 0.7 to 1.0% manganese, $\leq 0.1\%$ silicon, $\leq 0.075\%$ phosphorus and $\leq 0.008\%$ sulfur. For points and crossings subject to heavy traffic, the standard steel rails inevitably wore more quickly than elsewhere. The solution was to use manganese alloy steel, developed by R. F. Hadfield in 1883, with a composition of 12% manganese and 1.2% carbon. Its most famous application was in the complex junction just outside Newcastle Central Station, which was relaid with manganese alloy steel rails in 1912 (see Figure 7.1).

When Bessemer steel first became available in the mid-1860s it was used mainly for locomotive parts, but soon by far the largest demand for steel was for rails. Eventually each company employed

Figure 7.1 Part of the crossing outside Newcastle Central Station where the rails were made of the more durable manganese alloy steel.

a rails inspector, who travelled to the steelworks to check the quality of the rails before dispatch. He would have observed mechanical tests performed on a sample of the rails, inspected the chemical analysis performed in the works laboratory, and visually checked all the rails for surface defects.[14]

Most large railway companies had their own test house, where they performed independent mechanical tests, and they also sent samples of the rails to their own chemical laboratory for analysis. In a six month period in 1910, the North Eastern Railway laboratory received a total of 222 steel samples, of which 57 were of rail steel. All were analysed for the standard five elements (phosphorus, sulfur, carbon, silicon and manganese).[15]

In addition to analysing new rails, the chemists also tested steel rails that had broken. As early as 1873, before it appointed its own chemist, the Midland Railway found that the cause of a spate of broken rails was that the silicon content of a particular batch was too high.[16] In the period January 1888 to June 1891, Fuller, the chemist of the Lancashire and Yorkshire Railway, analysed 15 broken rails.[17] Between 1909 and the 1923 Grouping, the North Eastern Railway maintained a report book on broken rails, and in most cases a chemical analysis of the broken rail is quoted.[18]

By this time, however, it had been realised that factors other than an inappropriate chemical composition could be responsible for the breakage of rails and the failure of other steel components, and the railway chemists were involved in research on this issue (see Chapter 8).

As well as analysing new and broken rails, the chemist might be asked to analyse old steel rails which had been taken up, as these would be sold back to a steelworks for use in the open hearth furnaces. The laboratory might also analyse a variety of samples of steel intended for other applications, such as boiler plate, axles, tyres, brake blocks, drawhooks, steel rope, *etc*. As with rails, the candidates for analysis could be new items or old ones which had failed in service.

While performing these analyses, the railway chemists introduced improved methods for the determination of sulfur,[19] carbon[20] and manganese in steel. The much improved method for the estimation of manganese in steel provides one of the most important examples of the contribution to analytical chemistry made by the railway chemists. Reddrop at the Crewe laboratory reported that for some years prior to 1890 he had been evaluating the various methods available. He set out to improve one of the volumetric methods.[21] The method in question turned on the oxidation of manganese in solution to permanganate, followed by its estimation by titration with hydrogen peroxide. Reddrop, working with his colleague Ramage, found that the difficulties in working the method were caused by chloride present as an impurity in the reagents, and they devised a method for the preparation of hydrogen peroxide using a new solid oxidising agent, sodium bismuthate. The whole procedure could be accomplished in about 15 minutes.[22] The new method was the first rapid and reliable method for the determination of manganese in steel. Modifications were subsequently introduced by Brearley and Ibbotson,[23] and the method was used in this form until superseded by instrumental methods, being described in a well-known analytical text published in 1961.[24]

7.4.2 Copper

Of the non-ferrous metals, copper was the most widely employed in locomotive construction, being used for fireboxes, firebox stays and boiler tubes. As with many structural metals, small quantities of impurities can have a dramatic effect on its mechanical properties. Sulfur and oxygen were known to make copper brittle, and the railway chemists made improvements to the determination of both these elements in copper. In 1890, Phillips at Stratford East noted that there was an error inherent in the standard method for the estimation of sulfur in copper, which involved dissolving the sample in concentrated nitric acid and precipitating the sulfate formed as the barium salt. This gave a low result because barium sulfate is very slightly soluble in very acidic solutions. Phillips developed a procedure by which the majority of the acid was removed *before* precipitation, thus overcoming this problem.[25]

A few years later a paper appeared which maintained that the existing method for determining oxygen in copper was unreliable because it assumed all the oxygen was present as cuprous oxide; in practice it might be combined with another metal present as an impurity (*e.g.* lead).[26] Archbutt of the Midland Railway devised a method by which hydrogen was passed over the heated copper and the decrease in weight caused by the removal of oxygen as water vapour was measured.[27] A few years later, Archbutt's procedure was the only one described in a well-known manual of technical methods of chemical analysis.[28] Various metals, such as tin, lead and iron, were also thought to impair the properties of copper and were therefore frequently estimated. However, early in the twentieth century, research conducted by railway chemists revealed that traces of arsenic significantly improved the mechanical strength of copper, and thereafter railway companies specified that copper used in locomotive construction should contain small quantities of this element (see Chapter 8).

7.4.3 Other Metals, and Materials used in Foundry Work

The railway laboratories analysed samples of tin, lead, zinc, and alloys such as brass, bronze and white metal as required. The term white metal covers a range of relatively soft alloys which were widely used as bearing metals on railway wagons, *etc.* They were sometimes called "anti-attrition" or "anti-friction" metals. The Midland Railway white metal consisted of 84.2% tin, 10.5% antimony and 5.3% copper. In 1910, the North Eastern Railway had three different specifications for white metal. All contained less tin than the Midland specification, but also contained some lead, the amount varying in the range 4–17%. Every consignment was analysed on receipt, 19 samples being tested in the second half of 1910.[29] Furthermore, it was standard practice to melt and reuse old white metal, and sometimes these batches were analysed.

Although only the London and North Western and the Lancashire and Yorkshire companies possessed steel-making furnaces and steel foundries, all the large companies had iron and brass foundries. The chemists routinely analysed not only the metals to be cast in the foundries, but also the other materials used in foundry work. These included firebricks, fireclays, moulding sands, limestone, and a variety of ferro-alloys used either to adjust the composition of the melt or as deoxidisers. The chemists might also be asked to analyse the slags produced in the cupola furnaces of the iron foundries. One of the railway chemists devised an improved procedure for the estimation of the silica content of refractory materials.[30]

7.5 FUELS

Coal, a very complex and highly variable material, had largely displaced coke as a locomotive fuel before most of the railway laboratories were established (see Chapter 5). Coal was analysed for moisture; the coke it could yield (further analysed into "fixed" carbon and ash); and volatile matter. In addition, an elemental analysis in terms of carbon, hydrogen, nitrogen, oxygen and sulfur might be performed. A purchaser would want the carbon content to be as high as possible and the other substances to be present in as small a quantity as possible. The ash content of most coals was below 5%, but in some samples it was higher, and it was always a nuisance, not only because it lowered the calorific value and generated clinker, but it could also lead to deposits in locomotives known as "swallow's nests", at the junction of the boiler tubes with the firebox.[31] Sulfur could be present in coal in three forms: iron pyrites (FeS_2), combined organically with the carbon, and as calcium sulfate. As calcium sulfate it was harmless, but in the other two forms it was most undesirable. A problem associated with iron pyrites was that at high temperature it was oxidised to iron oxide, which could form a fusible slag with the silica of the ash. The bars in the firebox might thereby become covered, reducing the access of air and resulting in inefficient combustion of the coal. However, the main disadvantage of sulfur in coal resulted from the gases that were formed. Sulfur dioxide corroded the firebox and boiler tubes and polluted the atmosphere. Furthermore, when fresh coal was added to the fire, it was possible under certain conditions for the highly noxious gases hydrogen sulfide and carbon disulfide to be given off. Not surprisingly, these compounds, along with sulfur dioxide, caused particular problems in underground railways, and resulted in one enterprising London pharmacist selling a "Metropolitan Mixture" to ease the coughs of passengers.[32] Whatever the quality of the coal, however, pollution was always going to be a nuisance while steam locomotives were used on underground railways. The problem was only solved by the introduction of electric traction early in the twentieth century.

A vital characteristic of a fuel was its calorific value. A "theoretical" value could be calculated from the carbon, hydrogen and oxygen content of the coal,[33] but once suitable combustion calorimeters had been developed a better method was a direct calorimetric determination. Calorimeters of two designs were available and the first comparative evaluation of their performance in measuring the calorific values of coals was carried out by Reddrop at Crewe. He retired before he could publish the results, but they were presented on his behalf by Archbutt of the Midland Railway in 1901.[34] With the introduction of calorimetry, the railways relied less on assessing the value of a fuel by testing its steam raising power in a working situation, as they had done in the 1840s (see Chapter 5). However, similar experiments were still being performed at the end of the nineteenth century with the same fuel in different types of locomotive (*e.g.* standard and "compound") with a view to comparing their relative efficiencies.[35]

Some companies experimented with oil as a fuel for locomotives, and after the introduction of the internal combustion engine, a few petrol-driven railcars appeared.[36] In addition, the railway companies owned some petrol-driven road vehicles. Railway chemists used a piece of apparatus called the "bomb calorimeter" to determine the calorific value of liquid fuels. Most large railway works made their own coal gas, and Crewe made producer gas for the steelworks. Archbutt mentioned that he was analysing gaseous fuels in 1914,[37] and in 1920, Tipler stated that at Crewe he was analysing producer gas and coal gas, and also tar, ammonium sulfate and ferric oxide sent to him by the on-site gasworks.[38]

7.6 OILS FOR LIGHTING

The railway companies purchased "burning oils" in large quantities. These were used in the oil lamps situated behind the coloured glass lenses on semaphore signals, thus enabling them to be read in the dark. The chemists compared the illuminating power of various oils, and drew up specifications in terms of the relative quantities of the various fractions obtained on distillation. They also specified the flash point and specific gravity. By the end of the nineteenth century, mineral oils (*e.g.* kerosene) had largely displaced vegetable oils for use in signal lamps.

Oil lamps also provided the earliest form of carriage lighting. Initially rape oil was used but the light emitted was feeble. A huge improvement occurred from the mid 1870s when the railways began to use burners consuming oil gas, which provided a much brighter light. It had been known for some time that gas could be made from oil as well as coal. The basic process was to arrange for oil to drop into red hot retorts, where it decomposed into a gas with a relatively high illuminating power on account of the unsaturated hydrocarbons it contained. The railway companies manufactured the gas themselves, after which it was compressed and carried on trains in cylinders, usually underneath each carriage. The analysis of the oil (*i.e.* gas oil) purchased to make the gas was one of the responsibilities of a railway chemist. Indeed in 1891, the role of the Great Northern Railway's chemist, J. W Young, was widened to include the post of general superintendent of the oil gas works at London, Grantham, Doncaster and Leeds, with a working foreman under his charge at each location. For this additional responsibility he was granted a salary rise of £50.[39] The records for 1903 show that his successor, W. G. Young, analysed samples of gas oil delivered to the works each month.[40] Unfortunately, oil gas was the cause of some dreadful disasters when it ignited or exploded after sometimes minor collisions, and from the early twentieth century electric lighting became increasingly common.[41]

7.7 WATER

7.7.1 Water for Locomotives

In the early days of railways, water analyses were commissioned from consultants to measure the quantity of salts present that would form a deposit in the boiler, and also to indicate the presence of constituents that might cause priming (see Chapter 4). By the early twentieth century, large-scale automated water softening plants had become available, and railway companies began to install them. When a supply was being considered for softening, a complete analysis, giving the concentrations of all the substances present, was essential. This could then be used to calculate the quantities of the softening materials (lime and soda) that had to be added, and their cost. In 1905–6, the Great Northern Railway assessed all its water supplies in this way.[42] At the same time the relentless search for new or alternative supplies of acceptable quality continued. For example, the existing supply at Doncaster, the River Don, was judged to be poor, and the Great Northern was considering alternatives, with analyses supplied by the company's chemist.[43] Like the Great Northern, the Great Western Railway also conducted a survey of all its water supplies, performing a complete chemical analysis on each one.[44] It too was installing softening plants, and Churchward, the locomotive engineer, placed Bird, the chief chemist, in charge of them. In 1905, Bird provided Churchward with a number of suggestions for their more efficient management.[45]

The principal reason for using soft water in the boiler was that it reduced maintenance costs, but if hard water was used and the necessary washing out not performed, then a boiler explosion could result. An explosion on the Lancashire and Yorkshire Railway in 1906 was found to be caused by a layer of scale 3/8 inch which had been allowed to accumulate on the top of the firebox. The latter had become overheated and had softened, leading to its collapse and the subsequent explosion of the boiler.[46]

Another cause of boiler explosions was corrosion. All boilers corroded slowly from the inside, and ultimately the boiler plate could become so weakened that it would give way. A considerable amount of empirical evidence was accumulated prior to 1923 concerning the possible constituents of water most likely to cause corrosion,[47] and this provided another reason for obtaining a complete analysis of the water. However, at this time, the mechanism of corrosion was only partially understood. It is now established that a piece of iron or steel will contain irregularities in its surface, and as a result small galvanic cells can be set up which result in corrosion of the metal. The surface irregularities can be regions of different chemical composition, and can also be surface defects such as minute cracks. Cracks frequently arose when the rings of the boiler were constructed using lap joints (*i.e.* with the two pieces of plate overlapping). Such joints resulted in the boiler rings not being perfectly circular, but when subjected to the pressure of the steam, they tended to distort to a more circular shape, returning to their original shape when the boiler cooled. This continual flexing resulted in the formation of a minute crack parallel to the joint, which would then act as a centre for the occurrence of corrosion. The end result was the formation of a groove or furrow, and many boiler explosions were caused by the boiler plate giving way along such grooves. By the late 1860s it was known that grooving could be avoided by using butt joints instead of lap joints, but grooving caused by lap joints continued to cause explosions up until 1890 (see Figure 7.2).

Whatever the initial cause of the establishment of a galvanic cell, corrosion would occur more quickly if certain substances were dissolved in the water. It was well known that if the water was acidic then corrosion would ensue more rapidly, and even the acids found in peaty water could have a corrosive effect. Salts of iron,

Figure 7.2 An explosion of 1877, the result of corrosion along a groove near a lap joint in the boiler.[48] (National Railway Museum/SSPL).

aluminium, magnesium, and ammonium were known to decompose in the boiler to give acidic solutions and hence were undesirable in feed water. It was also found that nitrates and especially chlorides in the water assisted corrosion considerably. Hence the very detailed water analyses being provided by the railway chemists in the early twentieth century could be used to assess the likely corrosive effects of the water.

7.7.2 Water for Domestic Consumption

In addition, there was a greatly increased demand for analyses to be performed to assess the suitability of water supplies for domestic use. Railway towns such as Crewe and Swindon were virtually built by railway companies, who were initially responsible for their water supplies. However, most of the analyses of domestic water performed by the railway laboratories were done to assess small-scale supplies to stations, railway workers' cottages, *etc*. In the latter part of the nineteenth century there was considerable uncertainty and debate as to how water for domestic use should be analysed to see if it might cause disease and how the results should be interpreted. The "germ theory of disease" encountered opposition almost to the end of the nineteenth century, but there was the growing realisation that, in Phillips's words, the danger in water was of "living *animalculae* capable of propagating disease".[49] For many years the presence or absence of disease-causing agents was inferred by chemical tests. The rival chemical approaches were those of Frankland and James Alfred Wanklyn. The former argued that the nitrate content provided a measure of the "previous sewage contamination" in the water, whereas Wanklyn measured the albuminoid (*i.e.* proteinaceous) ammonia, which he argued measured the actual organic contamination.[50] Wanklyn maintained that in conjunction with a high chloride content, a high albuminoid ammonia content indicated contamination by animal waste, whereas a low chloride content indicated the contamination was of vegetable origin. A high free ammonia content, in conjunction with a low chloride content and a low or zero albuminoid ammonia content, indicated that the nitrogen was of mineral origin.[51] The Frankland method was more accurate, but also more difficult and lengthy than his rival's. A railway chemist often measured all parameters before forming his opinion. Thus when testing a well water in 1875, Reddrop quoted the total solids, the chlorine (*i.e.* chloride), temporary, permanent and total hardness, free ammonia, albuminoid ammonia, and nitrogen as nitrites or nitrates. He also examined the water microscopically, commenting that it contained "a considerable number of living organisms and animalculae", as well as a large quantity of suspended vegetable matter. He concluded that this water was "inadmissible for domestic use".[52] If a qualitative test on a domestic supply indicated the presence of a poisonous metal, such as lead, then its concentration would also be determined.

It was the development of bacterial culture methods by Robert Koch that led, in the period 1880–1900, to the isolation and identification of many disease-causing microorganisms.[53] It thus became clear that a thorough bacteriological examination was an essential component of an analysis of a water intended for domestic use. In the 1902 edition of his book, Phillips merely stated that when examining a water "it would be interesting to make, if possible, a microscopical search for any low forms of animal or vegetable life that it may contain".[54] However, a few years later it was recorded that the Swindon laboratory was "fully equipped with sterilizers, incubators and other accessories for bacteriological work on water".[55] Although today such investigations would be carried out by microbiologists, on the railways prior to 1923 this work was done by chemists.

7.8 LUBRICANTS

In the years leading up to World War I, the most important innovation in steam locomotive design was the introduction of "superheating". After collection from the boiler, the steam passed through small tubes contained within the flue tubes, thus raising its temperature before it was fed to the cylinders. A greater degree of expansion was thereby attained, and the steam passed through the feed pipes and cylinders without condensing, thus eliminating an important source of power loss. Superheating enabled locomotive power to be increased by as much as 25%. This engineering innovation could not be introduced, however, without the development of suitable lubricants for the cylinders, as traditional lubricants would have decomposed and carbonised under these more severe conditions.

At the coming of the railways, the lubricants used were all agricultural products, such as tallow oil, lard oil, neatsfoot oil, olive oil, rape oil and castor oil. These are all mixtures of glycerides (esters of glycerol), but in the terminology of the day they were known as "fixed" oils. Towards the end of the nineteenth century, mineral oils, which are mixtures of hydrocarbons, became available. Glycerides are prone to decomposition by high temperature water and steam, but hydrocarbons are not. It was the availability of mineral oils that enabled lubricants to be formulated that would perform satisfactorily in the cylinders of superheated locomotives. However, pure mineral oils are less satisfactory lubricants (see Chapter 8), so a small amount of fixed oil was often included in the cylinder oil for superheated locomotives. Formulating an oil which possessed satisfactory lubricating properties while retaining an adequate degree of stability was a difficult task, and the railway chemists advised on suitable mixtures. They also assisted in the formulation of oils for use in other situations, such as locomotive motions and axles. These latter oils could contain a higher proportion of fixed oil, as they did not come into direct contact with high pressure steam. As late as 1934, it was reported that the oil used on the *Flying Scotsman* contained a high percentage of olive oil, but that the cheaper mineral oils were used for shunting engines.[56]

In the early twentieth century, the most important physical measurements made on lubricating oils were the flash point, the specific gravity at various temperatures, the viscosity and the volatility. These parameters were the main ones used to decide whether or not a mineral oil was suitable for a certain application, and in the course of their work, the railway chemists devised and published a number of improvements to methods of measuring viscosity and volatility.[57] However if the sample was a fixed oil, or a mixture of fixed and mineral oils, then chemical tests were necessary. These determined the percentage of free acids, the proportions of mineral oil and fixed oil in a mixture, and the identity of an unknown fixed oil.

Before about 1880 very few methods were available for the analysis of fixed oils, and indeed the full extent of their complexity was not fully appreciated for many years. Taking olive oil as an example, although it consists principally of the oleic acid triester of glycerol, six other carboxylic acids are usually present, and the total number of possible combinations that can exist is therefore 196 (ref. 58). As a result, samples of olive oil from different sources will be slightly different, aside from the fact that they might also differ owing to the presence of contaminants. Possible contaminants included free fatty acids (which could be present if the oil had been left in contact with debris from the olives and decomposition, catalysed by enzymes, and had set in). Free fatty acids in a lubricating oil could attack the metal of the bearings, and the metallic soaps thus formed could cause the oil to become too thick, resulting in further damage. In addition, the oil producer might have employed some crude refining processes, such as treatment with concentrated acid, and traces of the acid may have been left in the oil. Possibly the most serious potential problem, however, was adulteration with a cheaper oil, and detection of such adulteration was difficult. Also, at this time, research was being performed on lubrication, with the goal of understanding how lubricants produce their effects and why some oils make better lubricants than others. Important work was done in this area by one of the railway laboratories in particular (see Chapter 8).

Free acids in oil were determined by a volumetric procedure, which was simplified by Archbutt.[59] In many instances, however, the railway chemists used, and sometimes adapted, methods which had recently been developed for the analysis of fats in foodstuffs. One such procedure was first introduced by Koettstorfer in 1879 for the examination of butter for foreign fats.[60] Koettstorfer described a method for the determination of the saponification value of the fat. Saponification is the decomposition (or hydrolysis) of the fat by alkali, so the saponification value is the quantity of alkali required to decompose unit mass of fat. The presence of foreign fats could be deduced from its numerical value, as butterfat contains a significant proportion of lighter short-chain fatty acids, and therefore requires more alkali per unit mass for saponification than lard, tallow, dripping and other possible adulterants. A similar procedure could be used to estimate the percentages of fixed and mineral oils in a mixture, as mineral oils, which consist of a mixture of hydrocarbons, have a saponification value of zero. Therefore if a lubricant was composed of a mixture of a mineral oil and a certain fixed oil (of known saponification value), the determination of the saponification value of the mixture enabled the percentage of the fixed oil to be calculated. The mineral oil could be separated by ether extraction after the saponification stage, the ether removed and the oil dried, weighed and its characteristics determined.[61] A variation on this procedure was introduced by Gripper, chief chemist at Gorton.[62]

A more difficult analytical problem was to identify an individual fixed oil or to detect the adulteration of one fixed oil by another. Although any fixed oil consists of a complex mixture of triglycerides, characteristic numerical values or "signatures" could be obtained for each oil (although these were narrow ranges rather than precise constants). One example is the saponification value, but for the fixed oils used in lubricants the saponification values do not differ greatly, so they are of limited use for identification purposes. However, a number of other signatures depend on the extent to which the fixed oil incorporates unsaturated fatty acids, and these values vary significantly between the various fixed oils used as lubricants. One of these signatures is the iodine value, a measure of the uptake of iodine by the oil, first introduced by Hübl in connection with food analysis in 1884.[63] Archbutt in 1886 was the first to point out the usefulness of the iodine value in the analysis of lubricants,[64] and later he reported that the most reproducible results were yielded by an improved method that had been introduced by Wijs.[65] Quicker procedures involved allowing the oil to react with bromine, either measuring the increase in weight on bromination (the bromo-gravimetric method),[66] or the accompanying temperature rise (the bromo-thermal method).[67] Jenkins published an evaluation of these methods when applied to lubricating oils,[68] and Archbutt commented that the bromo-thermal method was very rapid and provided a valuable additional method for the evaluation of lubricants.[69] Another characteristic value was the observed temperature rise when a certain quantity of oil was treated with a standard quantity of concentrated sulfuric acid, introduced in 1852 and known as the Maumené value. Its utility was greatly increased as a result of research over 30 years later by Archbutt, who found that the results were reliable and reproducible if the conditions were very carefully controlled.[70]

These were the most frequently used methods in the investigation of fixed oils used as lubricants, although a considerable number of additional tests were available and would no doubt have been used occasionally by the railway chemists. Of these miscellaneous methods, the only one improved by a railway chemist was the elaïdin test, which was a crude method of detecting the adulteration of olive oil. It was based on the fact that genuine olive oil is converted to a solid product by a reagent acting as a source of nitrous acid, whereas in the case of an adulterated olive oil the mixture remained fluid, or became only partially solid. Archbutt improved the test,[71] and in this form it had some application, but it was very soon superseded by more quantitative techniques, such as the determination of the iodine value.

It was the quantitative analysis of mixtures of fixed oils which provided one of the most difficult analytical problems for the railway chemists until well into the twentieth century. If two oils were thought to be present, the iodine value or another parameter would be determined, and the percentage of the two components deduced from the accepted values for the pure oils. In 1892, following analysis by their chemist, the Great Northern Railway fell into dispute with its supplier over the extent to which cottonseed oil was being added to the rape oil which they were purchasing for both lighting and lubricating purposes.[72] The Great Northern Railway's contract was for 10 000 tons of oil, so for the supplier much was at stake. Although it is not recorded which analytical methods were being used, it is very likely that the dispute arose because the different parties were employing different approaches. Initially the Great Northern Railway cancelled the contract, but relented when its chemist eventually obtained analytical results close to those obtained by the supplier.

Apart from oil, grease was used as a lubricant on the railways to an enormous extent. Grease was the lubricant of choice where retention of a fluid lubricant on the moving surfaces was a problem, and it was also significantly cheaper than oil. Grease, like oil, was originally applied by hand as required, but by the 1860s passenger carriages and some goods wagons were provided with axleboxes, which meant that the application of fresh lubricant was required at less frequent intervals. By the end of the nineteenth century the development of oil seals meant that most railway carriage axleboxes were oil filled, but grease was still frequently employed for lubricating wagons.

In the very early days of railways, tallow and lard were used as greases, but, by the middle of the nineteenth century, the leading railway companies were manufacturing grease to their own specifications. The basic process was to treat fixed fats or oils with a quantity of alkali dissolved in water, thus turning some of the oil into a soap that stabilised the remaining oil and water in the form of an emulsion. Each company had its own recipe, and as time went on, all modified their specifications to include some mineral oil and a certain amount of pre-prepared soap into the mixture. During 1903, W. G. Young of the Great Northern Railway analysed tallow for greasemaking from 602 casks supplied to the company, and he also analysed a further 29 samples before contracts were placed.[73] The companies' chemists were also involved in analysing the greases produced.

7.9 PAINTS

The railway companies used paints and varnishes on an enormous scale. Whilst the task of painting the Forth Bridge was the stuff of legend, thousands of smaller structures had also to be protected on a continuous basis from rust and decay. Other items such as locomotives, rolling stock, station buildings, *etc.*, required a coating with a high standard of decorative finish, as they were subject to close inspection by passengers, and their good appearance enhanced a company's reputation. Furthermore, the majority of items requiring a protective coating existed in a very hostile environment, being exposed not only to the weather on a continuous basis, but to attack from smoke, coal dust and general industrial pollution.

The paints and varnishes used up to 1923 were of traditional composition. Paints contained linseed oil as the vehicle. In the presence of air, linseed oil forms a cross-linked polymer which results in a tough but elastic skin. This process was known as "drying". The rate of drying was enhanced if the oil had been previously subjected to a procedure in which the oil was heated while air was forced through it, a process known as "boiling". The rate of drying could also be increased by incorporating one or more dryers into the paint. These were usually lead or manganese salts, which exerted a catalytic effect on the cross-linking process. The other major component of a paint was the body pigment, which gave the paint the necessary coating power and also conferred a colour. Commonly used body pigments included carbon (usually as lampblack), red lead, and iron oxide. When a decorative finish was required, a white pigment and a colour were sometimes employed.

Commonly employed white pigments were: white lead, barytes (barium sulfate), gypsum, whiting (chalk), kaolin and zinc white (zinc oxide). The paint made from these ingredients could be thinned with the addition of turpentine. Many railways mixed their own paints, and the chemists could be asked to perform a quality control analysis on any paint component. They also analysed premixed paints, varnishes and putty.

Of all the materials associated with the painting process that were investigated, turpentine presented the greatest challenge to the analyst. Turpentine is obtained by distilling sap which has been tapped from the trunks of various species of coniferous trees. The principal component of all turpentines is α-pinene, but the exact nature of turpentine from different sources varies considerably, depending upon the species of tree and the distillation technique employed. Good quality turpentine makes an excellent thinner for paint as it vaporises at ambient temperature leaving no more than a trace of residue. Turpentine could be adulterated with the more volatile fractions of petroleum. In the nineteenth century, such mixtures were regarded as being distinctly inferior to the genuine article, as the petroleum component left a greasy residue on evaporation.

A railway chemist would assess a sample of turpentine by measuring its specific gravity, refractive index, optical rotation, flash point, behaviour when subjected to fractional distillation and volatility. A chemical method for estimating the percentage of petroleum in a mixture was to treat a sample with moderately concentrated sulfuric acid or fuming nitric acid. The acids attack α-pinene, yielding a polymer in the case of sulfuric acid and water-soluble compounds in the case of nitric acid, but leaving the petroleum adulterants unaffected. Phillips (Great Eastern Railway) recommended the nitric acid method in 1891, and this was presumably the technique used in the Stratford laboratory.[74] However, when an organic compound is allowed to react with nitric acid, the products are often explosive, and indeed other workers later reported violent explosions when using this method.[75]

In the search for an alternative procedure, Archbutt (Midland Railway) experimented with the application of an iodine absorption technique. He found that the quantity of iodine absorbed by a sample of turpentine depended on the conditions, but as long as these were carefully controlled, reproducible results were obtained and it was possible to estimate the proportion of genuine turpentine in a mixture.[76] However, the method probably received only limited application, for by 1910, as a result of improved distillation techniques, the quality of light petroleum substitutes had improved, and Archbutt commented that there was no longer any serious objection to them.[77]

7.10 MISCELLANEOUS MATERIALS

The materials considered so far were purchased by the railway companies in huge quantities, and the routine analysis of samples of them was the major task of the chemists. In addition, the chemists might also be asked to analyse a large number of other materials on an occasional basis. A few examples will be quoted to illustrate the range of work required of a railway chemist.

In 1903, W. G. Young at the Great Northern Railway was testing metal polish, soft soap, disinfectant powder, jointing paste, window cleaning powder, enamel filler, glass lamp chimneys and ferro-aluminium (for use in the foundry).[78] Wallpaper was tested for arsenic content, because green wallpaper sometimes contained the pigment Scheele's Green (copper arsenite). It had recently been shown that under damp conditions mould on the wallpaper could release a gaseous arsenic compound into the air and cause low-level arsenic poisoning.[79] In 1910, E. W. Rowley of the North Eastern Railway was examining cloth (36 samples for railwaymen's uniforms), Iglodine and Septonal antiseptic dressings, Kyanizing liquid, and tar.[80] In 1916, the Great Western laboratory was assessing asphalt, creosote, lime, leather, glue and cement.[81]

Sometimes when a chemist analysed several samples of a similar product, the company was deciding who should be awarded a contract. In 1888, the chemist of the Lancashire and Yorkshire Railway, C. J. P. Fuller, analysed electroplated spoons before a contract was awarded by the refreshment rooms committee. Three manufacturers had submitted samples, and probably on the basis of the thickness of the plating, the committee awarded the contract to Mappin and Webb.[82] On one occasion a chemist was asked to decide on the relative merits of four samples of sherry, although the analytical criteria adopted in this case were not mentioned.[83] An analyst might demonstrate that a proprietary product had little or no advantage over a similar preparation produced in-house.

Thus, in 1903, W. G. Young analysed a commercial carriage cleaner and found it almost identical in composition to the Great Northern's own less expensive preparation. The cleaning staff had maintained that the proprietary product was superior.[84]

One material presented the chemists with problems to which they were unable to find satisfactory solutions. That material was rubber. The railways started to use rubber after 1852, when George Spencer (who had been an engineering draughtsman with a railway contacting firm) patented a rubber spring for use on railway carriages and wagons. In that year he formed his own company, and by 1890 it was supplying most of the large railways.[85] In 1893, the Lancashire and Yorkshire was running a trial to compare the use of rubber and laminated steel springs on carriages and wagons. As part of the trial, their chemist provided analyses of rubber springs from two manufacturers, but when reporting the results to the directors, Aspinall, the chief mechanical engineer, pointed out that it was difficult to obtain a reliable analysis of rubber.[86] As a result of such problems, it was very difficult for a chemist to draw up a specification in terms of percentage composition against which a supplied batch of rubber could be checked. The railway chemists attempted to improve the methods of rubber analysis, but in spite of these efforts they were unable to contribute much to the quality control of rubber goods.

7.11 ANALYSIS OF GOODS TENDERED FOR CARRIAGE, AND CLAIMS FOR DAMAGE

The railway chemist not only tested materials purchased by the company, but investigated a wide range of substances tendered for carriage. The railway companies operated what seems to modern eyes to be the most extraordinarily complex classification system of charging for the carriage of goods. This system was first instituted by the Railway Clearing House in 1847, and was subject to continuous revision (see Chapter 9). The companies were perpetually being presented with new materials for carriage, or goods for which the classification was not obvious from the description given when they were tendered. It was therefore necessary to decide to which class to assign the goods in order that the appropriate rate could be levied for their carriage, and the opinion of the chemist was often sought in these cases. As Archbutt commented in his 1914 address:

Questions are continually arising in connection with new kinds of traffic, such as their nature and value, the category into which they fall, whether they are provided for under the existing classification of goods, or need new and special provision to be made for them; questions also relating to the correct or incorrect description of goods handed to the railway for conveyance, and the justice or injustice of claims made for damage to goods during conveyance or storage.[87]

When a new material was offered for carriage, the chemist had not only to provide advice on classification, but also had to advise on any safety issues. In 1903, the Great Northern's chemist, W. G. Young, was asked to provide guidelines on "thermit, the new welding compound", and on "pneumatic tyre repair outfits containing inflammable liquids".[88] Such requests reflect the technological developments of the day, and in that year Young and his counterparts in other companies advised on many other new items presented for carriage. The chemist would give his advice for the consignment in question, but if it appeared that the new material might be tendered for carriage on a frequent basis in the future, the matter would be discussed by the railway chemists' committee at the Railway Clearing House (see Chapter 9). This committee might recommend that a new entry be added to the classification, as well as stipulating how a potentially dangerous material should be packaged, labelled and transported.

Not infrequently a consignor did not know into which category his goods fell, and gave his own description. In the second half of 1910, E. W. Rowley, the chemist of the North Eastern, was asked to rule on the classification of 35 such samples.[89] For example Rowley decided that a consignment described as "earth colour" should be classified as "paints and colours, class 2", and that a consignment described as "hat dye" should be classified as "spirit varnish".

With such a convoluted classification system, it is not surprising that consignors sometimes deliberately misdeclared goods in an attempt to benefit from lower rates. In 1903, Young found that a consignment of explosives had been declared as "dyes or colours", ultramarine had been declared as "lime blue", and petroleum as "lubricating oil".[90] Likewise in 1910, Rowley found that patent corn flour had been declared as "pea flour" and ferro-silicon had been declared as "ferro-silica".[91] Many other examples could be quoted. By detecting such misdeclarations, the chemist was able to prevent the company being defrauded by its customers, and where dangerous goods were being given a harmless description, the chemist was undoubtedly making a contribution to public safety.

As Archbutt indicated, the chemist might be called upon to investigate the validity of claims for damage. In 1903, W. G. Young described how he investigated 29 instances of damage to goods. In one case, the company was being held responsible by the consignor for damage to leather goods valued at £1000. Young was able to show that the damage had been caused by sea water, and therefore the company could not have been responsible. Several of the claims were from a sugar manufacturer who maintained that sacks of sugar were spoiling in the railway company's warehouses. Young found that the refiner had changed the manufacturing process by using sulfurous acid instead of using charcoal to clarify the sugar. Traces of the acid remained in the refined sugar, and over a period of time the slightly damp sugar was being decomposed by the acid. Young had therefore been able to show that the change in the refiner's process was the cause of the problem. In some cases a claim was found to be unrealistic. Thus, when a dog washing preparation called "Cutoxine" was damaged in transit, Young showed that the amount claimed was about seven times the value of the materials in the mixture.[92]

In 1911, Jenkins at the Great Eastern laboratory investigated several cases of damage to expensive pianos in transit. Sometimes the pianos acquired rusted wires or pegs, but more often the polished wooden surfaces became dull or milky. The sender sued the Great Eastern for damages. Jenkins suspected that the unlined deal cases in which the pianos were transported were losing moisture on warming, and that this moisture was condensing on the cooler surfaces of the pianos and causing the problems. He devised a series of experiments in which sample pieces of piano case were subjected to the same conditions of temperature and humidity as the pianos in transit (see Figure 7.3), and he found they acquired the same milky bloom as the pianos in transit. When this evidence was presented in court, judgement was given in favour of the railway company. The following year, the London and North Western experienced similar problems, and Jenkins's results were likewise presented as evidence when that case came to court.[93]

Figure 7.3 A page from the notebook of J. H. B. Jenkins (Great Eastern) detailing part of his study into the reason why piano cases acquired a milky bloom whilst in transit. (BRB (Residuary) Ltd.).

Sometimes a claim for compensation was fraudulent, and on one occasion a company brought a criminal prosecution, using the results of the chemist's investigation as evidence. A trader had made several claims against the London and North Western for the breakage of eggs in transit, so Tipler, the chemist, arranged for the next consignment of eggs to be invisibly marked. Tipler found that the shells of the broken eggs submitted by the trader did not show the marks, and the fraud was thereby exposed.[94]

There were of course cases where claims were justified. It was standard practice that when a number of small consignments of different goods were to be sent to the same destination, they would be loaded into the same wagon. As a result of poor packaging or rough shunting, it sometimes happened that bags or containers burst open and the materials from different consignments became intermixed and were spoiled. It was the chemist's job to investigate these cases, and to see if a portion of a contaminated consignment could be salvaged, thereby reducing the claim against the company.[95] Sometimes spoilage, if undetected, could have serious consequences (*e.g.* sugar being contaminated with weedkiller), and after such accidents the chief chemists of the various companies sitting in committee at the Railway Clearing House would make recommendations about appropriate safety measures to prevent a recurrence (see Chapter 9).

7.12 OTHER ACTIVITIES

From the foregoing, it is evident that the routine analytical work of the railway chemists was very wide ranging. Moreover, a railway chemist was expected to be on hand to use his scientific expertise whenever and wherever it was required. This is exemplified by a further quotation from Archbutt's speech of 1914:

> I used sometimes to be asked what a railway company could find for a chemist to do; the difficulty would be to answer in what direction of its numerous ramifications the railway company is unable to find work for the chemist. Even the Secretarial Department has consulted me in reference to the discovery of an indelible ink, and I am bound to add that up to the present I have been unable to find an ink which satisfies their requirements... When I was first appointed to my present position, some thirty years ago, there were very few railway chemists; and not even we ourselves, much less those appointing us, had any adequate idea of the work there was for us to do.[96]

The varied nature of these additional requests received by the chemist can be illustrated by three topics on which the Great Northern Railway's chemist provided advice in 1903. These were the extermination of rats in a goods warehouse; the prevention of weed growth in water troughs; and the cause of the growth of mould in policemens' helmets.[97] It is clear that the railway company chemist came to be regarded as a scientific jack-of-all-trades, and almost any scientific or technical issue which lay outside the competence of the engineers might be referred to him.

7.13 CONCLUSION

The materials testing service that the railway chemists provided was of great importance to their companies as well as being extremely varied in its scope. The principal benefit to a company was undoubtedly economic. With railway companies placing annual contracts for steel and oils in huge quantities, they simply could not afford to run the risk of inadvertently accepting faulty batches. Poor quality rails would very probably result in more frequent track repairs, and an inappropriate oil could lead to worn bearings. Likewise the use of an unacceptably hard water to supply locomotive boilers would inevitably lead to higher maintenance costs. Sub-standard paint would mean that repainting would be necessary more frequently. However, it was possible that the use of a sub-standard material could have consequences more serious than increased maintenance costs. Broken rails, fractured axles, seized bearings, *etc.* could result in breakdowns or even accidents, and an exploding boiler was almost always catastrophic. Aside from the deaths and injuries that accidents could cause, they were very expensive for the company, as indeed were breakdowns.

Intimately connected with the routine quality control work was the revision of specifications in the light of experience and as a necessary consequence of changes in the design of locomotives, *etc.* By their continuous revision of specifications, the chemists played an important role in the technological development of railways. In addition, the chemists' work in classifying materials tendered for carriage, investigating the causes of damage to goods, and detecting attempts to defraud the company, all helped to maximise railway company revenue.

In the course of their quality control work, the railway chemists made some valuable contributions to analytical chemistry. As important as this research was, however, it is overshadowed by the research the chemists did in helping to solve railway rather than chemical problems. It is to this railway orientated research that we must now turn.

REFERENCES

1. J. Simmons, *The Railway in England and Wales, 1830–1914*, Leicester University Press, Leicester, 1978, p. 231.
2. L. Archbutt, *Analyst (Cambridge U. K.)*, 1914, **39**(108), 113.
3. Work done by Analyst's Dept., GNR Reports to Board, 1902–3, Document 11 (TNA, RAIL 236/388).
4. An example is the white metal used in bearings, where the Midland and North Eastern had markedly different specifications.
5. An example of a paper discussing various possible compositions of rail steel is C. P. Sandberg, *J. Iron Steel Inst., London*, 1898, **2**, 76.
6. This committee was appointed to consider the strength of steel rails in the aftermath of the St. Neots accident of 1895. See Chapter 8.
7. These bodies represented the principal steel consumers and producers. The descendant of this committee is the British Standards Institution.
8. L. Archbutt and R. M. Deeley, *Lubrication and Lubricants*, Griffin, London, 1st edn, 1900, 5th edn, 1927.
9. Quoted in ref. 8, 3rd edn, 1912, pp. 141–6.
10. H. J. Phillips, *Engineering Chemistry*, Crosby Lockwood, London, 1st edn, 1891, 3rd edn, 1902.
11. Equivalent weights were experimentally determined combining weights – the weight of a material that would combine with one gram of hydrogen, eight grams of oxygen, or the equivalent of any other element or compound.
12. J. Reddrop, *Chem. News*, 1890, **61**, 245–250 and 256–258. See also J. Reddrop and H. Ramage, *J. Chem. Soc.*, 1895, **67**, 268–277 (footnote on p. 270). The normality method of expressing concentrations of solutions used in volumetric analysis had been in use for a number of years. The designation "E" was

intended to convey that the solutions were not standardised, unlike volumetric solutions designated by "N".
13. R. Blochmann, *Ber. Dtsch. Chem. Ges.*, 1890, **23**, 31.
14. An engaging account of the work of one rail inspector is to be found in C. J. Allen, *Two Million Miles of Train Travel*, Ian Allan, London, 1965, pp. 25–49.
15. NER Chemical Laboratory, Half Yearly Report, 31.12.1910 (TNA, RAIL 527/1372).
16. Minute 182, 04.06.1873, MR Board of Directors, 1873–6 (TNA, RAIL 491/22).
17. L&YR Committee Minutes, 1888–1891 (TNA, RAIL 343/310, p. 354; RAIL 342/312, pp. 105, 323, 359; RAIL 343/314, pp. 198, 200, 471, 587; RAIL 343/316, pp. 522, 523; RAIL 343/320, p. 502; RAIL 343/322, pp. 156, 288, 414, 497; RAIL 343/324, pp. 142, 454; RAIL 343/326, pp. 185, 544).
18. NER Reports on Broken Rails, 1909–1923 (TNA, RAIL 527/1238).
19. L. Archbutt, *J. Soc. Chem. Ind., London*, 1890, **9**, 25.
20. B. Blount and A. G. Levy, *Analyst (Cambridge U. K.)*, 1909, **34**, 88. (See pp. 96–97).
21. L. Schneider, *Dinglers Polytech. J.*, 1888, **269**, 224; *Idem*, *J. Soc. Chem. Ind., London*, 1888, **7**, 693.
22. J. Reddrop and H. Ramage, *J. Chem. Soc.*, 1895, **67**, 268.
23. F. Ibbotson and H. Brearley, *Chem. News*, 1901, **84**, 247.
24. A. I. Vogel, *A Textbook of Quantitative Inorganic Analysis*, Longmans, London, 3rd edn, 1961, p. 299. One of the authors [CAR] has vivid personal memories of using this method in an industrial laboratory before instrumentation superseded it.
25. H. J. Phillips, *Chem. News*, 1890, **62**, 239.
26. B. Blount, *Analyst (Cambridge, U. K.)*, 1896, **21**, 57.
27. L. Archbutt, *Analyst (Cambridge, U. K.)*, 1900, **25**, 253; *Idem*, 1905, **30**, 385.
28. G. Lunge and C. A. Keane, *Technical Methods of Chemical Analysis*, Gurney and Jackson, London, vol II, pt.1, 1911, p. 194.
29. Ref. 15.
30. L. Archbutt, *J. Soc. Chem. Ind., London*, 1892, **11**, 215.
31. *Locomotive Magazine*, 1905, **11**, 82.
32. Ref. 1, p. 130.
33. See *e.g.* ref. 10, 3rd edn, 1902, pp. 138–45.

34. Archbutt presented Reddrop's results during the discussion of a paper by A. Adams, *J. Soc. Chem. Ind., London*, 1901, **20**, 1084.
35. Results of a series of experiments performed on the Great Eastern in 1886 are given in Ref. 10, 3rd edn, 1902, p. 175.
36. An example is provided by the London Brighton and South Coast Railway, which introduced some petrol railcars in 1905.
37. Ref. 2.
38. F. C. Tipler, Typescript of lecture, *Work of a Railway Chemist*, delivered to the Association of Crewe Locomotive Works Foremen, 12.03.1920. Author's collection.
39. Minute of 24.02.1891, GNR Locomotive Committee, 1890–1893 (TNA, RAIL 236/200).
40. Ref. 3.
41. Ref. 1, p. 202.
42. GNR Details of Qualities and Quantities of Water for Locomotives, with Maps of Locations of Water Supplies, 1906 (TNA, RAIL, 236/690).
43. GNR Doncaster, Water Supply for Locomotives, 1905–1906 (TNA, RAIL 783/80).
44. GWR Laboratory Notebook, "Waters", 17.3.1904–11.07.1905 (WSRO, Acc No. 2515).
45. Letter W. R. Bird to G. J. Churchward, 19.01.1905. Copy in ref. 44.
46. C. H. Hewison, *Locomotive Boiler Explosions*, David and Charles, Newton Abbot, 1983, pp. 112–3.
47. P. G. Jackson, *Boiler Feed Water*, Griffin, London, 1st edn, 1919. Jackson had been a chemist in the Derby laboratory before moving to the National Boiler and General Insurance Company Ltd. in 1908.
48. Another cause of boiler explosions was the failure of a group of firebox stays (see Chapter 8).
49. Ref. 10, 1st edn, 1891.
50. For the methods of Frankland and Wanklyn, see C. A. Russell, *Edward Frankland, Chemistry, Controversy and Conspiracy in Victorian England*, CUP, Cambridge, 1996; C. Hamlin, *A Science of Impurity, Water Analysis in Nineteenth Century Britain*, Hilger, Bristol, 1990; C. A. Russell in, *Instruments and Experimentation in the History of Chemistry*, ed. T. H. Levere, MIT Press, Cambridge, MA, 2000.

51. J. A. Wanklyn and E. T. Chapman, *Water Analysis*, Kegan Paul, Trench, Trübner and Co., London, 10th edn, 1896, pp. 70–1.
52. LNWR Register of Water Analyses, 1853–1883, entry of 07.11.1875 (CCA, NPR/3906).
53. W. Bulloch, *The History of Bacteriology*, OUP, Oxford, 1938.
54. Ref. 10, 3rd edn, 1902, p. 237.
55. A. J. L. White, *GWR Magazine*, 1916, **28**, 163.
56. J. I. Hill, *London and North Eastern Railway Magazine*, June 1934, 330.
57. Viscosity: ref 10, 3rd edn, 1902, p. 243b; F. Lidstone, *J. Soc. Chem. Ind., London*, 1917, **36**, 270; *idem, Phil. Mag.*, 1922, **43**, 354; *idem*, 1922, **44**, 953. Volatility: L. Archbutt, *J. Soc. Chem. Ind., London*, 1896, **15**, 326.
58. For x fatty acids combining with glycerol, the total number of possible compounds is $(x^2+x^3)/2$. R. J. Taylor and C. H. S. Hitchcock, *The Chemistry of Glycerides*, Unilever, London, 1982.
59. L. Archbutt, *Analyst (Cambridge, U. K.)*, 1884, **9**, 170.
60. J. Koettstorfer, *Z. Anal. Chem.*, 1879, **18**, 199; *idem, Analyst (Cambridge, U. K.)*, 1879, **4**, 106.
61. A. H. Allen, *Commercial Organic Analysis*, Churchill, London, 2nd edn, vol. 2, 1886, p. 83.
62. H. Gripper, *Chem. News*, 1892, **75**, 27; *Idem, J. Soc. Chem. Ind., London*, 1892, **11**, 182.
63. A Hübl, *J. Soc. Chem. Ind., London*, 1884, **3**, 641.
64. L. Archbutt, *J. Soc. Chem. Ind., London*, 1886, **5**, 303, pp. 308–9.
65. L. Archbutt, *J. Soc. Chem. Ind., London*, 1904, **23**, 306. For the Wijs method, see J. J. A. Wijs, *Ber. Dtsch. Chem. Ges.*, 1898, 750.
66. O. Hehner, *Analyst (Cambridge U. K.)*, 1895, **20**, 49.
67. O. Hehner and C. A. Mitchell, *Analyst (Cambridge, U. K.)*, 1895, **20**, 146.
68. J. H. B. Jenkins, *J. Soc. Chem. Ind., London*, 1897, **16**, 193.
69. L. Archbutt, *J. Soc. Chem. Ind., London*, 1897, **16**, 309.
70. E. J. Maumené, *C. R. Hebd. Seances Acad. Sci.*, 1852, **35**, 572–3; L. Archbutt, ref. 64, pp. 303–4.
71. L. Archbutt, ref. 64, pp. 304–308.
72. Minutes of 05.10.1892, 02.11.1892, 22.11.1892, 21.02.1893, 03.05.1893, GNR Locomotive Committee, 1890–1893 and 1893–1896 (TNA, RAIL 236/200 and 201).

73. Ref. 3.
74. Ref. 10, 1st edn, 1891.
75. J. H. Coste, *Analyst, (Cambridge, U. K.)*, 1908, **33**, 219, p. 231.
76. T. F. Harvey, *J. Soc. Chem. Ind., London*, 1902, **21**, 1437; *idem, ibid.*, 1904, **23**, 413. Archbutt's observations are on pp. 1439 and 414 respectively.
77. J. H. Coste, *Analyst (Cambridge, U. K.)*, 1910, **35**, 112–117 (Archbutt's comments are on p. 115).
78. Ref. 3.
79. J. C. Whorton, *The Arsenic Century: How Victorian Britain was Poisoned at Home, Work, and Play*, OUP, Oxford 2010.
80. Ref. 15.
81. A. J. L. White, *GWR Magazine*, 1916, **28**, 163.
82. Minute of 20.03.1888, L&YR Refreshment Rooms Committee, Committee Minutes 21.12.1887–08.05.1888, p. 441 (TNA, RAIL 343/310).
83. Anonymous account of the history of the NER laboratory, probably prepared at the request of T. H. Turner (chief chemist, LNER), *ca.*1932 (NRM/DA, 245).
84. Ref. 3.
85. P. L. Payne, *Rubber and Railways in the Nineteenth Century: A Study of the Spencer Papers*, Liverpool UP, Liverpool, 1961.
86. Minute of L&YR Directors, 29.11.1893, Committee Minutes 25.10.1893–28.03.1894, p. 127 (TNA, RAIL 343/338).
87. Ref. 2.
88. Ref. 3.
89. Ref. 15.
90. Ref. 3.
91. Ref. 15.
92. Ref. 3.
93. J. H. B. Jenkins, Personal Photograph Album and Notebook, 1904–1920, pp. 31–3 and 36, (NRM/DA, 575).
94. Ref. 38.
95. S. Wise, in *British Railway Research – The First Hundred Years*, ed. A. O. Gilchrist, Institute of Railway Studies, York, 2000, p. 8.
96. Ref. 2.
97. Ref. 3.

CHAPTER 8

Research on Railway Issues

8.1 INTRODUCTION

The materials testing role of the railway chemists has been described in the previous chapter. This relatively routine function also involved investigative work of various kinds, such as improving analytical methods; solving problems associated with the classification of items tendered for carriage; investigating the spoilage of goods; and countering attempts to defraud the company. But by the 1890s, some of the railway chemistry laboratories were sufficiently well equipped, and employed a sufficient number of chemists, to undertake research which contributed directly to the technological development of the railways. Before this work is discussed, however, we must consider the context in which it was undertaken.

8.2 THE BACKGROUND TO RAILWAY RESEARCH

In the late nineteenth century, a dedicated research department was a rarity in British industry. As Donnelly has commented with reference to analytical laboratories in the chemical industry of the period:

...a significantly subsidiary process was the expansion of laboratory activities into forms of embryonic research and development... Often employees created the opportunities to undertake such activity by a kind of internal scientific entrepreneurship, viewed with varying degrees of suspicion, neutrality or approval by employers.[1]

Although these remarks refer to the chemical industry, the research conducted by both chemists and engineers employed by the railway companies prior to 1923 also fits this description. After the Grouping one of the new companies, the London Midland and Scottish Railway, created a formal research organisation, and in 1932 its director (the chemist, Sir Harold Hartley, FRS) made some revealing comments about railway research in the pre-Grouping days. In a paper read to the Institute of Transport, he described the history of railway research, outlining what had been achieved before World War I. However, this had been accomplished with little overt official encouragement from the management:

...for many years before the war original work of permanent value had been done in railway laboratories and workshops. It would be true to say, however, that these early investigations were made despite the lack of any definite research policy laid down by the Boards of the railway companies, and were mainly the pioneer enquiries of individuals.[2]

Hartley went on to quote examples of "original work of permanent value" done by both engineers and chemists prior to Grouping. Among the engineering projects were those of Churchward (Great Western) on locomotive valve gears and on the circulation of water in the boiler, and the work of Sir Henry Fowler (Midland) on the best material for locomotive crank axles and on the development of nickel steel chisels. The major chemical research projects were centred on metals, lubrication, and water treatment, but in addition the chemists were involved in a significant number of minor projects.

8.3 METALS

Although other materials were tried, copper, with its excellent thermal conductivity, was the most commonly used material for

Figure 8.1 The interior of a locomotive firebox. The tube plate is at the rear and the riveted heads of the stays can be seen at the sides. (National Railway Museum/SSPL).

fireboxes, firebox stays, and boiler tubes (see Figure 8.1). In the early days of railways, copper as supplied had good mechanical properties, but ironically difficulties began to occur in the 1880s because improved refining processes for copper were being introduced. This change was in part stimulated by the demands of the infant electrical industry. The world's first public electricity supply was provided in Godalming in 1881, and the new electrical industry wanted the purest possible copper with the lowest possible resistance. As a result, commercial copper was becoming too pure and hence too soft for constructional purposes. The new soft copper was resulting in greatly increased maintenance costs for the railway companies as fireboxes were being abraded by the fuel, and the tube plates were cracking. The latter problem was so expensive to repair that in 1884, the Midland built ten "1667" class engines with the number of tubes (and hence the number of holes in the tube plate) reduced from 223 to 205, but because of the reduced heating surface the locomotives were less powerful. The alternative of increasing the diameter of the boiler and retaining the standard number of tubes would have resulted in a heavier engine.[3]

The chemists were called upon to ascertain which impurities conferred the desired properties on copper, and in what quantities.

Initially the approach was somewhat *ad hoc*. Railway chemists were sometimes sent samples of firebox plate by the chief mechanical engineer to ascertain the differences between copper from fireboxes that had worn well and from those that had performed poorly. For example, in 1891 Phillips (Great Eastern Railway) quoted analyses he had performed on copper taken from a 40-year-old firebox which had given good service and a much more recent firebox which had failed. He measured the percentages of 11 elements in the two samples, and his conclusion was highly tentative: '...what is the best hardening ingredient to add to copper so as to reduce corrosion and abrasion to a minimum, has yet to be determined; although arsenic in limited quantities would seem a favourable addition for that purpose.'[4]

Two years later, William Dean, locomotive superintendent of the Great Western Railway, gave a paper on tensile tests and chemical analyses of copper plates removed from fireboxes which had been taken out of service after varying mileages. Dean quoted analyses of plates taken from 18 locomotives, and the percentage of each of 14 foreign elements in the samples was estimated. The analyses were carried out by the company's chemist F. W. Harris between December 1884 and April 1893.[5] Dean's conclusion was even more cautious than that of Phillips: '...it would probably be unsafe to form from these tests any detailed conclusion as to the best composition for copper fire-box plates. It would appear however that the presence of a small percentage of arsenic is in no way detrimental to the lasting quality of the plates.'

The attempt to understand the effect of an impurity on the mechanical properties of a metal was a hot research topic of the day. The leading academic worker in this field was William Roberts-Austen, professor of metallurgy at the Royal School of Mines. Immediately prior to Dean reading his paper to the Institution of Mechanical Engineers, Roberts-Austen read a paper mainly concerned with alloys in which copper was the principal component. He tested the mechanical properties of specially prepared mixtures. His principal conclusions were that a small percentage of arsenic seemed to result in improved mechanical properties, and that bismuth had an injurious effect. In an appendix, Roberts-Austen alluded to the durability of copper fireboxes. He quoted the analysis of copper from a firebox of a locomotive of the Metropolitan Railway which performed satisfactorily for over 20 years and had covered 500 000 miles.[6] The principal impurities found were lead, arsenic and nickel.

Tentative as these conclusions were, railway companies began to specify that arsenical copper should be used for fireboxes and boiler tubes. However, a few years later Archbutt of the Midland Railways cast some doubt on Roberts-Austen's conclusion concerning the harmful effect of bismuth. Roberts-Austen's work had resulted in engineering specifications stating that copper should contain a maximum of 0.005% of this element. Archbutt subsequently reported that during the period 1891 to 1902 he analysed 100 samples of copper firebox plate from various makers and found percentages of bismuth ranging from nil to 0.046%, the average being 0.015%. The fireboxes performed satisfactorily, and Archbutt concluded that some of the other impurities present in commercial copper must neutralise the injurious effects of the bismuth. Archbutt then pursued this matter in conjunction with E. A. Lewis, a metallurgist from Birmingham, who initially worked on the matter in his own laboratory but subsequently spent 12 months working with Archbutt at Derby.[7] They tested samples of copper (prepared from pure electrolytic copper to which impurities had been added) by performing mechanical tests in the laboratory. They rolled the copper samples into sheets at red heat, hammered them to assess their forging properties, and conducted bending tests on cold annealed samples. Not only did they confirm the beneficial effect of arsenic, but they concluded that arsenic renders harmless small quantities of lead and bismuth. For this reason, and because its tensile strength is higher than refined copper, they suggested that arsenical copper was one of the best materials for locomotive fireboxes and tubes.[8] In 1927, E. L. Ahrons, in his classic history of the development of the British steam railway locomotive up to 1925, wrote 'The pure copper was by no means so suitable for fireboxes as that generally supplied in former years. It was subsequently found that the beneficent impurity in the copper of old fireboxes was arsenic, and, as a result, it is now specified as obligatory'.[9] However, Ahrons made no mention of the part that the chemists had played in demonstrating the vital role of arsenic.

Although the properties of the copper used for fireboxes and boiler tubes was of great importance from a maintenance point of view, from a safety point of view the durability of the stays was of even greater significance. The firebox was surrounded by boiler water under pressure, and the stays pinned the firebox to the boiler casing. Should a group of stays fail, the wall of the firebox would burst inwards, and when the boiling water came into contact with the red hot coal a catastrophic explosion would ensue (see Figure 8.2). As the locomotive evolved, boiler pressures increased from 50 pounds per square inch in 1830 to as much as 200 pounds per square inch in the early twentieth century. It probably seemed obvious to some that the strongest possible stays should be provided, but research at Crewe published in 1902 showed that this was not the case.

The work at Crewe was supervised by Tipler, the chief chemist of the London and North Western, but the subsequent paper appeared under the authorship of F. W. Webb, the chief mechanical engineer.[10] The Tipler/Webb experiments involved fitting locomotives with stays made from a variety of materials. As well as observing how well each kind of stay performed in

Figure 8.2 The result of an explosion at Westerfield near Ipswich in 1900. The left hand side of the firebox had been blown off the stays, causing the boiler water to come into contact with the red hot coal. (National Railway Museum/SSPL).

locomotives, detailed chemical analyses were performed on each material and tensile strength tests were performed over a range of temperatures up to 750 °F. Finally, Webb and Tipler performed riveting tests in which they found that copper stays formed the best riveted joint in a copper firebox. Furthermore, they reached the very important conclusion that the mechanical properties of the stay itself were not as important as had hitherto been imagined, and whatever the composition of a stay, it should be softer than the copper firebox into which it was riveted. They found that an excellent joint was formed between copper firebox plate containing 0.55% arsenic and a copper stay containing 0.34% arsenic.[11] The disaster depicted in Figure 8.2 occurred because the stays were made of bronze, and their joint with the softer firebox was weakened as the latter expanded and contracted.

Two years later, Webb and Tipler followed up their work on stays with a set of experiments on the durability of boiler tubes made from various copper alloys. They tested tubes from ten different makers, putting either 19 or 20 of each make of tube into the same engine. Unlike the work on stays, the tubes were not deliberately manufactured to have different compositions; there was a sufficient variation in the composition of the copper supplied by the various manufacturers. When a tube failed, it was replaced and a chemical analysis performed. After the engine had completed 142 348 miles, the experiment was terminated, and an average chemical analysis (for 15 elements) was reported for tubes from each group that had not failed. Webb and Tipler concluded that 'the most satisfactory copper tubes for locomotives are those containing either 3 per cent nickel, or at least 0.5 per cent of arsenic'.[12]

In the early days of the use of steel for rails, a rail sometimes broke because the composition of the steel was inappropriate. But by the 1890s it was becoming clear that a rail could fail even though its carbon and manganese content were within specification, and its silicon, sulfur and phosphorus content were below the accepted maximum values. At the end of the nineteenth century, metallography, a technique for the investigation of the microcrystalline structures of metals, started to come into general use. Metallography was soon to be used to explain the causes of rail failures, the first occasion coming in the wake of a fatal accident at St. Neots in 1895 when a rail broke under the weight of the locomotive hauling an express (see Figure 8.3).[13]

Research on Railway Issues   135

Figure 8.3   Sketch of the aftermath of the St. Neots derailment of 1895. (*Illustrated London News*).

The accident was the subject of an enquiry by the Railway Inspectorate of the Board of Trade, which came to the conclusion that the rail had broken because it had deteriorated since it had been laid.[14] The findings of the enquiry spurred the Board of Trade to appoint a committee in 1896 with the terms of reference:

> ...to enquire what extent loss of strength in steel rails is produced by their prolonged use on Railways under varying conditions, and what steps can be taken to prevent the risk of Accidents through such loss of strength.

The members of the committee, which delivered its report in 1900, included Roberts-Austen, Edward Thorpe (principal chemist of the government laboratory), and the steel magnate, Lowthian Bell.[15] Thorpe conducted chemical analyses on the St. Neots rail and a number of other broken rails. He found that in no case, including that of the St. Neots rail, did the chemical analysis show a composition which would have given cause for alarm.

Roberts-Austen then examined the rail using the relatively new technique of metallography, or "photomicrography" as he called it. This involved polishing and etching the surface of a section of the metal, and then photographing it as seen though a microscope.

Roberts-Austen found that the St. Neots rail contained patches of "martensite", a brittle form of steel produced by quenching from a high temperature. He disagreed with the opinion of the Railway Inspectorate that the rail had deteriorated in service, and concluded that it had been brittle from the start. He made some suggestions as to how the rail manufacturing process could be altered to ensure that such a rail could not be produced in the future. In one of the appendices to the report, Roberts-Austen and Thorpe said "The question of the mode of existence of carbon in the iron, and its general distribution in the rail... is not touched upon by the Railway Companies, who do not as yet appear to have adopted photomicrographic analysis." The railway companies were being given the message that they should examine their new and broken rails not only by chemical analysis, but by metallographic methods as well.

The larger railway chemistry laboratories adopted metallography almost immediately, as is evidenced by publications emanating from them. In 1905, Jenkins, (Great Eastern Railway), with his assistant Riddick, published a paper in *The Analyst* entitled "The microscopic examination of metals".[16] This gave a very clear account of Jenkins's work with the new technique, illustrated by many photographs of samples of steel and copper, including some of faulty components such as a fractured mild steel shaft (see Figure 8.4). Although these were not the first such photographs to be published, they made an important contribution, for as one of the metallurgists present at the reading of the paper commented, there was an enormous amount of learning to be done. The railway chemistry laboratories thereby contributed to the body of knowledge which enabled the detection of microcrystalline structures which could lead to the failure of rails and other components. Metallography was thereafter used by the railway laboratories as a tool in their quality control work, but the technique was soon employed in other research projects.

In 1905, Archbutt of the Midland Railway published photomicrographs demonstrating how the microstructure of copper was affected by small amounts of oxygen,[17] and he conducted research into how the microstructure of the white metal alloy used in bearings varied with its composition, particularly the percentage of zinc.[18] Archbutt also used metallography to study a problem first brought to his attention in 1908 by the Midland's telegraph

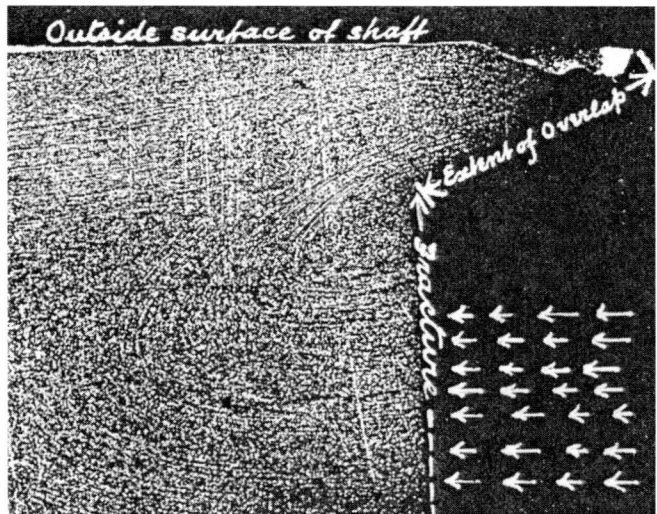

Figure 8.4 A photograph at five-fold magnification of part of a longitudinal section of a fractured mild steel shaft after polishing and etching, taken by Jenkins and Riddick at the Great Eastern laboratory in 1905. The flaw which caused the fracture is clearly seen.

superintendent. It had been found that some of the lead sheathing used to protect telegraph cables was developing cracks and failing after only a few years, but in other cases the sheathing was still perfectly satisfactory after 40 years. Over the following years, Archbutt was sent further samples of sheathing that had failed on the Midland, and he also received samples from the power superintendent of London Electric Railways and the chief electrical engineer of the London and North Western Railway.

It was not until 1921 that Archbutt was able to publish on the causes of the problem.[19] Chemical analysis had shown nothing unusual in the sheathing that had cracked, and there was no significant difference in the composition of the lead between sound and defective regions of the same length of sheathing. However, metallography revealed that in the portions which had failed, the metallic crystals were larger than in the sound regions, and that the cracks followed the course of the boundaries between the crystals. Archbutt then conducted stress and vibration tests on samples of lead, and found that continued stress caused the crystals gradually to lose their cohesion and come apart. He also found that the unusual structure in the region of failure occurred at the point when the hydraulic press, which extruded the lead sheathing on to the cable, had been stopped to recharge it. Archbutt suggested that the cable makers review their procedures to try to prevent the problem arising. He concluded by mentioning that an alloy of lead containing 3% tin, which had a much finer microstructure and was less likely to undergo intercrystalline failure, was under trial on the Midland for sheathing cables. After Grouping (and after Archbutt's retirement), these experiments were continued under the auspices of the British Non-Ferrous Metals Research Association, and resulted in the formulation of an alloy that found wide application.[20]

8.4 LUBRICANTS AND LUBRICATION

The introduction of mineral oils as lubricants in the second half of the nineteenth century has already been discussed (see Chapter 7). Their stability compared to fixed oils made them ideal for lubricating steam cylinders, as fixed oils were prone to decomposition by steam with the formation of acids which could attack the metal of the cylinders. Mineral oils were especially useful in lubricating the cylinders of superheated locomotives, where the operating temperature was significantly higher. However, it was found in practice that the lubricating power of mineral oils was less than that of fixed oils of similar viscosity, so compound oils consisting of a mineral oil mixed with a small percentage of a fixed oil such as lard oil were formulated for such applications. The observed difference in lubricating power between the two types of oil remained unexplained, however, until the early twentieth century when the problem was attacked by both chemists and mechanical engineers.

A very important partnership between a railway chemist and a railway engineer was that between Archbutt and Deeley (Figures 8.5 and 8.6), respectively chief chemist and (eventually) locomotive superintendent of the Midland Railway. Although Deeley had no chemical qualifications, he displayed a considerable interest in chemistry. In the 1890s, when there was considerable speculation about the causes of the periodicity of the chemical elements, he published six papers in the chemical literature on the periodic law. As result of their collaboration, Archbutt and Deeley produced a

Figure 8.5 L. Archbutt.

Figure 8.6 R. M. Deeley. (W. W. Winter, Derby).

book, *Lubrication and Lubricants* (see Figure 8.7), which ran to five editions and became a standard text.[21] Furthermore, they each conducted their own research on various aspects of lubrication.

> # LUBRICATION AND LUBRICANTS.
>
> A TREATISE ON THE
>
> ## THEORY AND PRACTICE OF LUBRICATION,
>
> AND ON THE
>
> ### NATURE, PROPERTIES, AND TESTING OF LUBRICANTS.
>
> BY
>
> LEONARD ARCHBUTT, F.I.C., F.C.S.,
> CHEMIST TO THE MIDLAND RAILWAY COMPANY;
>
> AND
>
> R. MOUNTFORD DEELEY, M.I.Mech.E., F.G.S.,
> LOCOMOTIVE SUPERINTENDENT, MIDLAND RAILWAY.
>
>
>
> **SECOND EDITION, THOROUGHLY REVISED AND ENLARGED.**
>
> LONDON:
> CHARLES GRIFFIN AND COMPANY, LIMITED,
> EXETER STREET, STRAND.
> 1907.

Figure 8.7 Title page of *Lubrication and Lubricants.*

Chemists dealing with lubricants wanted to learn as much as possible about their composition. In 1901, Archbutt conducted a piece of research into the make-up of rosin grease used for lubricating wagons.[22] This was prepared using rosin oil, which was obtained from the destructive distillation of the resinous by-product formed in the production of turpentine. Another project involved graphited lubricants ("Aquadag" and "Oildag"), which first became available early in the twentieth century following the invention of a method for preparing colloidal graphite. In 1917, the newly formed Department of Scientific and Industrial Research (DSIR) set up a committee to study the whole question of lubrication and lubricants, with both Archbutt and Deeley as members. A report on the graphite lubricants, containing analyses performed by Archbutt, was published separately from the main report of the committee.[23] Another railway chemist who studied lubricants was Gripper (Great Central Railway). Fixed oils could slowly change if they were stored for long periods, and the nature of such changes was a matter of interest and concern to the railways. In 1899, Gripper published a study of some old samples of rape oil, ranging in age from between four and ten years, finding that the oils had undergone oxidation. Gripper speculated on the molecular changes occurring on oxidation, but a full understanding of the process lay some years in the future.[24]

As well as studying the composition of lubricants, railway scientists also contributed to theoretical studies on lubrication. The early work on lubrication theory was carried out in academia by Tower and by Reynolds.[25] These studies established that a well-lubricated shaft becomes separated from its bearing by a thin film of oil, which is under pressure. Under these conditions, the frictional resistance is due solely to the viscosity of the oil, and its chemical nature is immaterial to its effectiveness as a lubricant. Although maintaining an adequate film of oil between the parts in relative motion was the ideal, in practice this was not always achievable. Conditions where an inadequate film of oil existed included high load, low speed, and starting from rest. Lubrication under such conditions became known as boundary (or solid film) lubrication, as opposed to viscous (or liquid film) lubrication. Under boundary conditions it was realised that some property other than the viscosity of the oil was responsible for its effectiveness as a lubricant, for different oils of the same viscosity were found to vary in their lubricating ability. Lubricants which performed well under boundary conditions were said to possess a high degree of "oiliness", although there was initially no understanding of the mechanism of this kind of lubrication.

One of the problems addressed by the DSIR lubrication committee soon after it was established was that of oiliness and boundary lubrication.[26] Mirroring the experience with cylinder oils, it had long been known that in the boundary condition the effectiveness of a mineral lubricant was increased by the addition of some animal or vegetable oil. Deeley designed a piece of apparatus to compare the oiliness of various pure lubricants, and in 1918 he reported his results to the committee. His apparatus measured the coefficient of friction between two metal surfaces in contact and under pressure (*i.e.* the boundary condition) when immersed in a bath of oil. He found that when animal and vegetable oils were in the bath, the coefficient of friction was less than with mineral oils of the same viscosity, thereby confirming that the former possessed the property of oiliness to a superior degree.[27]

Archbutt discovered that the component of animal or vegetable oils responsible for their superior oiliness was the small percentage of free fatty acid which they usually contained, and that by adding 1–2% of a fatty acid to a mineral oil its properties as a lubricant were improved considerably.[28] The view that free acid in a lubricant was an evil to be avoided (because of the corrosion it might cause) therefore had to be modified, and it became accepted that a small percentage of free fatty acid in a mineral oil resulted in a lubricant as good as that produced by the addition of a larger percentage of fixed oil. However, Archbutt's work was pre-empted by Wells and Southcombe who patented the idea,[29] and by 1927 they claimed that for many applications the new oils had largely displaced the compounded oils made by mixing mineral oils with 10–20% fixed oils.[30]

Shortly before this work was published, Irving Langmuir in the United States commenced what would come to be seen as pathbreaking research on surface films on liquids and solids He concluded that, on a water surface, fatty acids form a monomolecular layer with their functional groups interacting with the water and their hydrocarbon chains pointing away from the surface. He also suggested that similar monomolecular layers could be adsorbed on to the surfaces of solids.[31] The study of surface phenomena was one of Langmuir's major research interests, and was to result in the award of the Nobel Prize for chemistry in 1932.

Langmuir's work of 1916–7 did not refer specifically to the subject of lubrication, but in 1920 he described an experiment in which he coated a glass surface with a layer of oleic acid one molecule thick. This had a dramatic effect on the slipperiness of the surface, and was very difficult to remove.[32] Whilst Langmuir is credited with explaining the property of oiliness in terms of a monomolecular film attached to the surface, Deeley had described a similar mechanism at a stage when he was unaware of Langmuir's work. Thus he stated "It would appear that the unsaturated molecules of the lubricant enter into a firm physico-chemical union with the metallic surfaces, thus forming a friction surface, which is a compound of oil and metal".[33] The separate contributions of Deeley and Langmuir were both very important, as they provided a theoretical explanation for the empirical observation that small quantities of additives can dramatically affect the properties of a lubricant.

8.5 WATER TREATMENT

In some areas no soft water for locomotives was available, and where the water was very hard, companies began to consider the option of softening the water. It was not until the latter part of the nineteenth century that suitable water softening plants for railway use first became available. A number of different designs were marketed, one of which was developed and patented by Archbutt and Deeley of the Midland.

Railway locomotives consumed huge quantities of water; a watering station at a moderately busy location would need to supply around 500 000 gallons per day. A chemical method of softening water was patented by Thomas Clark in 1841, which involved the addition of the quantity of lime water calculated to precipitate the (temporary) hardness.[34] Clark envisaged that the precipitate would be separated by sedimentation or by filtration, but both these methods presented problems. If the very fine precipitate was allowed to sediment naturally, the process was slow, requiring many hours, and large settling tanks had to be built. Filtration on a large scale was also difficult. Furthermore, the process was only really suitable if the water was temporarily hard because of dissolved calcium bicarbonate. Magnesium, originally present in the form of magnesium bicarbonate, was only partially precipitated owing to the appreciable solubility of magnesium carbonate, and permanent hardness in the form of calcium sulfate was not remedied at all. The ideal softening plant was one which could produce an adequate supply, could remove permanent as well as temporary hardness, could be operated more or less continuously, and was inexpensive to install and operate. The first improvement on Clark's process was introduced by J. H. Porter in 1876, when he employed a cloth filter press to remove the precipitate. By 1884, he had automated the process, the method being known by the name "Porter-Clark".[35]

In the early 1880s, filtration plants were devised using lime and soda in combination, which had the effect of precipitating both the permanent and temporary hardness. The mixture also enabled the magnesium to be more completely precipitated as magnesium hydroxide by forming caustic alkali *in situ*. Alum was often added to aid coagulation of the precipitate. Plants working on these principles were patented by Maignen and Gaillet, and by Huet.[36] Once these improvements in water softening technology had been introduced, railway companies started to install softening plants.

There was still room for further improvement, because all these softeners suffered from the disadvantage that the filters needed repeated attention. The ideal was a lime-soda apparatus which dispensed with the need for filtration altogether and operated on a continuous basis. Archbutt and Deeley developed a design, patented in 1891, which achieved these objectives. They devised a process by which the sludge left behind at the bottom of the tank from the previous softening operation was stirred by a steam jet to mix with the current batch being treated. This had the effect of seeding the mixture, which meant that precipitation occurred faster and the particles were coarser. As a result, precipitation and sedimentation were complete in about 30 minutes. By having three tanks of moderate size in operation in sequence (treatment, sedimentation and drawing off), an almost continuous stream of softened water was produced. The first plant was installed at Derby, and had an output of 30 000 gallons of softened water per hour (see Figure 8.8).[37]

Several railway companies, in addition to the Midland, installed Archbutt and Deeley plants. One such was the Great Central, which installed them at Annesley and Tuxford.[38] Another was the Great Western, which by 1906 was operating one at Fox's Wood supplying the water troughs at Keynsham, near Bristol.[39] The installation of a softening plant involved what would be called in modern parlance "a cost–benefit analysis", and the chemists were frequently asked to calculate the cost of softening individual supplies. The number of softening plants on the railways slowly increased, but the policy of using soft water almost exclusively was only adopted after Grouping.

Figure 8.8 The Archbutt and Deeley water softening plant at Derby in 1898, showing the three tanks which enabled a continuous supply of soft water to be maintained. (Reproduced by kind permission of Professional Engineering Publishing/SAGE Publications Ltd., London, Los Angeles, New Delhi, Singapore and Washington DC, from L. Archbutt, Water Softening and Purification by the Archbutt-Deeley Process, *Proceedings of the Institution of Mechanical Engineers*, 1898, pp. 404–54, copyright of Professional Engineering Publishing/SAGE Publications Ltd.).

8.6 MINOR RESEARCH PROJECTS

In addition to their work on metals, lubricants and water, the railway chemists were also involved in a number of other research projects. Many of these involved matters that were internal to their particular company and did not result in any kind of publication. Most of the company records that would have contained details of these projects have not survived, but enough is extant to provide a few examples of this kind of work.

In 1912, Tipler of the London and North Western Railway described a project conducted by his laboratory which resulted in a considerable saving for the company. This was the modification of the oil lamps used in railway semaphore signals to give better performance. Tipler described this work when he was making a case to the chief mechanical engineer for improved laboratory accommodation:

> Had we followed the lead of the other large railway companies in this country, we should have scrapped all the signal lamps at

present in use, the estimated value for which is £18,000 and in addition would have had to pay £34,000 to purchase a new complete patent lamp, that requires costly oil for seven days' burning. The successful work carried out in the laboratory resulted in the adaptation of our present lamp at an estimated cost for alteration of £8,300 giving a lamp that burns very satisfactorily and continuously for 14 days and at the same time burns a much cheaper oil, and on further investigation we constructed a lamp that will burn without attention for 28 days, and 34 have been under test for the last 12 months.[40]

This is an example of a piece of "internal" research which did not result in publication, but which was of very considerable benefit to the company.

During the period covered by this study, the railway companies purchased premixed paints and also mixed their own. The chemists were frequently called upon to conduct investigations on paints. A problem with a paint arose in 1909 when the Great Eastern found that when "MacPherson's Engine Blue" was used and then afterwards varnished, the varnished surface gathered into a series of lines on drying. Jenkins found that the paint had been made by grinding up the paste with linseed oil which contained 20% of a thin mineral oil. A series of experiments revealed that, although paints prepared in this way appeared to dry well, the undesired effect was produced on varnishing whenever the mineral oil was used in the paint and was independent of the nature of the pigment.[41]

For decorative coatings, white lead generally made an excellent base pigment, with another colour added if required, but the paint could be discoloured by a highly polluted atmosphere. In such circumstances, zinc white was preferable, with barytes and whiting being cheaper alternatives. In 1903, W. G. Young, chief chemist of the Great Northern, was conducting experiments to see which of the varieties of zinc white paint on the market gave the best results in practice.[42] Between 1916 and 1918, Jenkins was conducting experiments on the weathering of strips coated with paints containing six different white pigments.[43]

It is in the nature of research that not all projects have successful outcomes. For example, in 1897, Japanese wood oil (otherwise known as "tung oil") first appeared on the European market,

13 000 tons being exported from China and Japan in that year alone. Its drying properties were superior to those of linseed oil, leading to speculation that it would replace linseed oil in paints. Jenkins of the Great Eastern conducted experiments, publishing two papers on the chemical characteristics of the oil.[44] The early promise of the oil was not fulfilled, and by 1909 it was reported that it could not replace linseed oil.[45] In 1915, Jenkins conducted experiments on luminous paint for station name boards. The composition of the paint is not given, but the luminous (in reality phosphorescent) paints of the period contained components such as "good luminous calcium sulfide".[46] The phosphorescence was due to the heavy metal impurities (*e.g.* bismuth) which these ingredients contained.[47] The results of Jenkins's experiments were not promising.[48] Chemists concerned with lubricants were always on the lookout for new oils that might be superior in lubrication performance or lower in price than those in current use. For example in 1898, Archbutt published an investigation into curcas oil, obtained from a nut grown in the tropics, and in the following year he reported results on maize oil.[49] In both cases he found the oil to be semi-drying (*i.e.* its viscosity increased on exposure to air), and therefore unsuitable as a lubricant.

In the early years of the 20th century, oil gas (as opposed to coal gas) was widely used for lighting passenger trains (see Chapter 7). When the gas was compressed, a liquid (known as "oil gas hydrocarbon") was deposited in the compressors and became a subject of interest from both theoretical and practical points of view. From the standpoint of theory, there was a desire to understand the nature of the changes that had taken place to the original oil during the pyrolysis process and, from the practical point of view, the railway companies were interested to know if this by-product that they were generating had any useful applications. By 1920, oil gas was being slowly phased out for carriage lighting, but it was still used for cooking in restaurant cars, and the total annual production of the oil gas hydrocarbon by-product was 100 000 gallons. Lewis-Dale, a chemist at the London and North Western, conducted research at the Crewe laboratory on the liquid condensed from the plant which was then in use by the company. The two main objectives of the work were to see if the composition of the liquid produced was substantially the same as that reported previously by non-railway chemists studying the liquid from a different kind of plant, and to see if any use could be found for the liquid. Lewis-Dale came to the conclusion that the liquid was practically the same as that examined by the earlier workers. Among the possible applications considered for the liquid were converting it into motor fuel and using it to prepare a synthetic rubber. However, both these were judged to be impracticable. The research resulted in a thesis which was approved by the University of London for the award of the degree of PhD in 1924, and in the publication of a paper based on the findings.[50]

8.7 CONCLUSION

A great many engineers contributed to the enormous technological development which occurred in the railway industry between 1830 and 1923. Working alongside the engineers, however, were the chemists, and in certain critical areas their contribution to this development was vital. It is a contribution which up until now has not been adequately recognised.

REFERENCES

1. J. Donnelly, *Br. J. Hist. Sci.*, 1991, **24**, 3.
2. H. Hartley, *J. Inst. Transport*, 1931–2, **13**, 495.
3. E. L. Ahrons, *The British Steam Railway Locomotive, 1825–1925*, Ian Allan, London, 1963, pp. 269–70.
4. H. J. Phillips, *Engineering Chemistry*, Crosby Lockwood, London, 1st edn, 1891, pp. 38–9.
5. W. Dean, *Proc. Inst. Mech. Eng.*, 1893, 139. The internal report of the first analysis, performed in December 1884, and signed by Harris, has survived (NRM/DA, 630).
6. W. C. Roberts-Austen, *Proc. Inst. Mech. Eng.*, 1893, 102.
7. MR Locomotive Dept. Salary Book No. 2, 1897–1907 (TNA, RAIL 491/1070).
8. E. A. Lewis, *Engineering*, 1903, **76**, 753. See also L. Archbutt, *J. Inst. Metals*, 1912, **7**, 262.
9. Ref. 3.
10. F. W. Webb, *Minutes of Proceedings of the Institution of Civil Engineers*, 1902, **150**, 87. Webb stated (p. 114) that Tipler "had had charge of the experiments".
11. Ref. 10, pp. 103–5.

12. F. W. Webb, *Minutes of Proceedings of the Institution of Civil Engineers*, 1904, **155**, 401.
13. Anon, *Illustrated London News*, 1895 (16 November), **107**, 595.
14. A contributory factor was the heavy axle loading of the Stirling "single" locomotive hauling the train. After a similar derailment a few months later, locomotives of this class were rebuilt to lower the axle loading. L. T. C. Rolt, *Red for Danger*, Pan, London, 3rd edn, 1978, pp. 86–7.
15. Parliamentary Accounts and Papers, 1900, **30**(1), *Railways*.
16. J. H. B. Jenkins and D. G. Riddick, *Analyst (Cambridge, U. K.)*, 1905, **30**, 2.
17. L. Archbutt, *Analyst (Cambridge, U. K.)*, 1905, **30**, 385.
18. L. Archbutt, *J. Inst. Metals*, 1912, **7**, 266.
19. L. Archbutt, *Trans. Faraday Soc.*, 1921, **17**, 22.
20. Ref. 2, p. 497.
21. L. Archbutt and R. M. Deeley, *Lubrication and Lubricants*, Griffin, London, 1st edn, 1900; 5th edn, 1927.
22. L. Archbutt, *J. Soc. Chem. Ind., London*, 1901, **20**, 1193.
23. DSIR Bulletin No. 4, *Report on Solid Lubricants*, HMSO, London, 1920.
24. H. Gripper, *J. Soc. Chem. Ind., London*, 1899, **18**, 342.
25. B. Tower, *Proc. Inst. Mech. Eng.*, 1883, 632; *idem, ibid.*, 1885, 58; O. Reynolds, *Philos. Trans. R. Soc. London*, 1886, **177**, 157.
26. DSIR, *Report of Lubrication Committee*, HMSO, London, 1920.
27. R. M. Deeley, *Proc. Phys. Soc. London*, 1919–20, **32**.
28. H. E. Wells and J. E. Southcombe, *J. Soc. Chem. Ind., London*, 1920, **39**, 51T. Archbutt described his experiments during the discussion of this paper (p. 55T).
29. H. E. Wells and J. E. Southcombe, *Br. Pat.*, 1 303 77 (1918).
30. Ref. 21, 5th edn, p. 293.
31. I. Langmuir, *J. Am. Chem. Soc.*, 1917, **39**, 1848.
32. I. Langmuir, *Trans. Faraday Soc.*, 1920, **15**, 62.
33. R. M. Deeley, ref. 27, p. 10s.
34. T. Clark, *Repertory of Patent Inventions*, 1841, **16** (New Series), 225. The chemical equation for the process is $Ca(HCO_3)_2 + Ca(OH)_2 \rightarrow 2CaCO_3 + 2H_2O$.
35. J. H. Porter, *J. Soc. Chem. Ind., London*, 1884, **3**, 51.
36. B. Latham, *J. R. Soc. Arts*, 1884, **32**, 920; P. A. Maignen, *J. Soc. Chem. Ind., London*, 1886, **5**, 223.

37. L. Archbutt and R. M. Deeley, *J. Soc. Chem. Ind., London*, 1891, **10**, 511; L. Archbutt, *Proc. Inst. Mech. Eng.*, 1898, 404.
38. 'Loco', *Great Central Railway Journal*, 1915–6, **11**, 54.
39. W. R. Bird, *Transactions of the GWR Mechanics' Institution Junior Engineering Society*, 1905–6, 109.
40. R. E Jones, *Brief History of Crewe Laboratory*, unpublished typescript, 1961 (NRM/DA, 69).
41. J. H. B. Jenkins, Personal Photograph Album and Notebook (NRM/DA, 575).
42. Work done by Analyst's Dept., GNR Reports to Board, 1902–3, Document 11 (TNA, RAIL 236/388).
43. Ref. 41.
44. J. H. B. Jenkins, *J. Soc. Chem. Ind.*, London, 1897, **16**, 195; *Analyst (Cambridge U. K.)*, 1898, **23**, 113–8.
45. J. Lewkowitsch, *Chemical Technology and Analysis of Oils, Fats, and Waxes*, Macmillan, London, 4th edn, 1909, vol. 2, p. 65.
46. L. E. Andes, *Oil Colours and Printers' Inks*, Scott Greenwood, London, 1903, pp. 156–62.
47. J. R. Partington, *A Textbook of Inorganic Chemistry*, Macmillan, London, 6th edn, 1950, p. 784.
48. Ref. 41.
49. L. Archbutt, *J. Soc. Chem. Ind., London*, 1898, **17**, 1009; *idem, ibid.*, 1899, **18**, 346.
50. P. Lewis-Dale, PhD thesis, University of London, 1924. *idem, J. Soc. Chem. Ind., London*, 1925, **44**, 189T.

CHAPTER 9

The Railway Chemists in Collaboration

9.1 COLLABORATION OR COMPETITION?

Over certain routes, two or more railway companies entered into fierce competition, the most famous example being the rivalry which developed towards the end of the nineteenth century between the London and North Western, the Midland, and the Great Northern Railways for the lucrative Anglo–Scottish traffic. The Midland tempted passengers to patronise its longer and slower route over the Settle–Carlisle line by introducing in 1881 smoother riding eight and twelve-wheeled coaches and by providing restaurant cars, whereas its two rivals competed in terms of higher speeds and hence shorter journey times. This latter rivalry culminated in the famous "races to the north" of 1888 and 1895.[1]

However, whilst there may have been bitter commercial competition between companies, on a professional level their senior employees such as the chief mechanical engineers gathered together at meetings of the Institution of Mechanical Engineers. A similar situation pertains today, when for example chemists from rival pharmaceutical companies might meet together at a conference organised by the Royal Society of Chemistry. From 1892, the chief

---

Early Railway Chemistry and its Legacy
By Colin A. Russell and John A. Hudson
© Colin A. Russell and John A. Hudson 2012
Published by the Royal Society of Chemistry, www.rsc.org

chemists of the various companies were meeting together at an organisation known as the Railway Clearing House in London. Their deliberations mostly centred round the important issue of the safe transport of dangerous goods.

9.2 THE RAILWAY CLEARING HOUSE AND THE CARRIAGE OF DANGEROUS GOODS

In spite of sometimes fierce competition, the railway companies had been working together on matters of mutual interest for 50 years prior to 1892. As soon as the railways had spread across the land to the extent that it was possible to make lengthy cross-country journeys using more than one company, it was realised that some form of collaboration was essential. As a result the Railway Clearing House was established in 1842, initially to organise the through booking of passengers; to divide passenger receipts on a mileage basis; to encourage through transport of goods on a rate per mile basis; and to provide for the settlement of all inter-company debts.[2] Soon the Clearing House was dealing with many other issues, such as recommending that all companies adopt Greenwich Mean Time; attempting to standardise signalling systems; and discussing a range of safety issues.

The issue that was eventually to involve the chemists was first discussed at the Clearing House as early as 1847, when a common categorisation of goods was agreed. While companies were free to set their own rates for carriage of goods, it was realised that division of revenues would be simplified if a common system of classification were adopted. This classification became progressively more elaborate and complicated. By 1879 it filled a 129-page book.[3]

In legal terms railways were classed as "common carriers", which meant that they were obliged to carry the goods of anyone willing to pay their charges. However, the Railway Clauses Act of 1845 exempted them from carrying any goods that they considered to be dangerous. Nevertheless the railways' industrial customers became increasingly insistent in their demands that potentially hazardous materials be carried, and from 1855 the railway companies began to allow some dangerous materials to be transported, initially by special agreement for each consignment. In 1865, the Clearing House formulated a set of regulations covering various categories of dangerous and explosive materials. In 1875 the Explosives Act was passed, a clause of which required the railway companies to make by-laws governing the carriage of explosive materials. These by-laws adopted a classification (proposed by the government) of explosives into seven categories, and specified appropriate regulations for materials in each class. Further legislation followed in 1888, with the passage of the Railway and Canal Traffic Act. This required the railways to prepare a revised classification of goods of all descriptions for government approval, which was regarded as being a necessary step towards a fairer and more uniform system of charging. In addition to the new classification, it was decided that a special classification should be created for explosives and other dangerous goods. The first draft of the special classification was drawn up by a sub-committee consisting of the goods managers of five companies (known as the "explosives committee"), who then asked their company chemists to comment on their proposals. The chief chemists from the companies concerned did not meet as a group at this stage; they each sent their comments to their respective goods managers. Previous regulations covering the carriage of dangerous materials had been made by the goods managers alone, but now the stage had been reached when they felt the need for specialist advice, and their company chemists were in possession of that necessary expertise.

The newly created special classification was much more than a charging structure for dangerous goods, for it contained "conditions of carriage". These specified how each material should be packed and labelled, how it should be stored, and any special precautions which should be taken when it was being loaded or unloaded. To finalise the special classification, it was decided that some of the issues should be further discussed at meetings of the goods managers' explosives committee and their company chemists. These meetings, which took place in early 1892, are the first occasions known when chemists from different companies met together. The final agreed special classification retained the seven classes of explosives of the government classification of 1875, but added to these were three categories of other goods of a dangerous nature: inflammable liquids, dangerous or corrosive chemicals, and (inevitably) miscellaneous.[4]

The leading figure amongst the railway chemists in these deliberations was H. J. Phillips of the Great Eastern. Four years later, as a result of his involvement in framing the special classification, he published a book entitled *The Handling of Dangerous Goods*.[5] This contained, in non-technical language, a description of the properties of all the common dangerous and explosive chemicals likely to be encountered. It also contained a lengthy section entitled "Instructive Accidents" which recounted, sometimes in lurid detail, various disasters that had occurred in recent years, not all of which had occurred in connection with the railways. The book concluded with a reprint of the new special classification. It was, in effect, an early safety manual. It is impossible to assess its impact, but it probably made a valuable contribution to public safety.

9.3 THE RAILWAY CHEMISTS' COMMITTEE

Although a group of railway chemists had met with the goods managers on several occasions in 1892, there was as yet no intention of establishing a standing committee of railway chemists. However, it was only a matter of months before the goods managers' explosives committee needed further expert chemical advice. An explosion and fire at the London goods depot of the Midland in March 1893 was caused by the chemical sodium peroxide. It was only recently that this material had been manufactured and transported in large quantities, and on this occasion a canister was faulty, resulting in leakage of sodium peroxide which led to the explosion. Leonard Archbutt, the chemist of the Midland Railway, submitted to the committee two designs for canisters which would be much safer.[6] Only nine weeks later, another accident, this time on the Lancashire and Yorkshire at Bradford Exchange station, resulted in Archbutt again being consulted. A fourteen year-old lad was carrying two cylinders of compressed gas, one containing oxygen and the other coal gas, through the station subway, when the oxygen cylinder slipped from the boy's shoulder, exploding when it hit the ground. This was one of the accidents Phillips subsequently described in his book in his characteristically gory fashion: "The unfortunate boy was killed instantly by the flying metal, his head being shattered and his body mutilated." The breakage was caused by the steel being of too high a tensile strength, and hence too brittle. Archbutt attended the meeting of the explosives committee when this matter was discussed.

The following year (1894), the goods managers' explosives committee discussed the specification of a drum for the carriage of ether. This very inflammable liquid vaporises very easily, and hence a considerable pressure can build up inside a drum which is left in hot sunshine. The explosives committee decided they should hold a joint meeting with the chemists from three companies (Midland, Great Western and Great Eastern), and when this was inconclusive, it was decided that the chemists should meet alone. The same three chemists, along with the chemist of the Great Northern, met at the Railway Clearing House on 22 May 1894 to consider this issue. This is the first known meeting at which a group of railway chemists from different companies met alone at the Railway Clearing House. They concluded their deliberations at a second meeting, at which they also considered various other issues at the request of the goods managers, such as the classification of rubber solution and regulations for the conveyance of nitric acid.

A pattern now became established that, with minor modifications, was to continue until Grouping (and beyond). The explosives committee of goods managers referred questions to the chemists, and the chemists' minutes on these issues were then considered by the explosives committee at a subsequent meeting. The explosives committee, which nearly always accepted the chemists' recommendations, duly reported back to the full meeting of goods managers, which in turn accepted the chemists' advice. Initially, the explosives committee appointed a special meeting of chemists on each occasion they deemed it necessary, but from the beginning of 1897 the chemists were meeting on a regular basis without any specific instructions to do so.

Over the years, the role of the chemists' committee increased in importance. In 1894, the explosives committee met for a total of 11 days, and the chemists met for three, with the two groups holding one joint meeting. In 1909, the dangerous goods committee (as the explosives committee had been renamed) met for four days, the chemists for 16 days. Furthermore, since 1895 there had always been some chemists in attendance at the meetings of the explosives/dangerous goods committee. It is therefore quite clear that the railway chemists moved from a position prior to 1892 of having no input at the Clearing House, to a position a few years later when they were the dominant partners in the deliberations on dangerous goods. In the early days of the railways, the goods managers probably considered themselves adequately qualified to deal with all issues concerned with the transport of dangerous goods, but from 1892 they began to realise that the involvement of chemists, with their specialised knowledge, was essential.

Although nine companies possessed chemistry laboratories in 1892, only four (the Great Eastern, the Midland, the Great Western and the Great Northern) were represented by their chemists at the meetings which took place with the goods managers that year. However, by 1916 all the companies which possessed laboratories (with the exception of the Great North of Scotland, which remained outside the Railway Clearing House) were sending a representative to the meetings of the railway chemists' committee (see Figure. 9.1).

Figure 9.1 The chief chemists of the railway companies at the Clearing House around 1915. Bird (Great Western) is in the chair, with Archbutt (Midland Railway) to his left. Behind Archbutt is Jenkins (Great Eastern Railway), and to his left is Tipler (London and North Western Railway). (BRB (Residuary) Ltd.).

9.4 NEW WORK FOR THE COMMITTEE

The special classification of 1892, which the chemists had helped to draw up, was in constant need of amendment and augmentation. New materials were being presented for carriage on a frequent basis, and the chemists were called upon to assess any likely hazards they possessed. An example from 1909 concerns nitrate of lime fertiliser. This material was beginning to come into the country in large quantities from Norway, where it was produced by treating lime with nitric acid. This manufacture was a spin-off from a new process for the manufacture of nitric aid, which utilised the abundant hydroelectric power available in Norway. Since there was no entry in the classification for nitrate of lime, it was necessary to assess if there were any dangers associated with it, and in particular whether it should be placed in the special classification along with nitrate of soda (a dangerous material). In examples such as this, one of the chemists would conduct the necessary investigation in his own laboratory, and report his findings back to the committee. In this case, the experiments were conducted by Gripper of the Great Central, and, as a result of his work, the chemists recommended that the conditions which applied to nitrate of soda were unnecessary for nitrate of lime. They advised that it should not be listed in the special classification of dangerous goods; all that was necessary was a new entry in the general classification. The volume of work associated with the examination of new materials steadily increased: for example, in 1912, a total of 68 such investigations was parcelled out among members of the committee.

One of the most dangerous materials carried by the railways was petroleum spirit. From the mid-nineteenth century onwards the quantities of petroleum products that needed to be transported increased enormously. At first, the trade was mainly in paraffin oil for lighting purposes, but with the invention of the internal combustion engine, the railways started to transport grades of petroleum suitable for use as a motor fuel. Prior to the establishment of the chemists' committee, there had already been a number of accidents involving petroleum, the most serious being in 1866 at Abergele in North Wales. A collision occurred between the Irish Mail and some runaway wagons speeding down an incline towards it. The rear wagons were carrying 50 barrels of paraffin oil, and as a result of the impact they burst open, and their contents were set ablaze by the spilt coals from the locomotive firebox. In the ensuing fire, 33 people were burned to death. In 1901, with the increasing quantities of petroleum being transported, it was agreed that specially constructed tank wagons were preferable to large numbers of small barrels loaded into ordinary trucks. The chemists, working with their engineering colleagues and H. M. Inspector of Explosives, were involved in designing a suitable tank wagon. It was also agreed that when smaller quantities had to be transported, specially constructed drums should be used, rather than ordinary barrels. These regulations were clearly successful, for, when eight years later, the Home Office reviewed the situation, the chemists were able to report that there had been no accidents involving petroleum on the railways since the regulations came into force, except in cases where those regulations had been infringed.

Another matter which occupied the committee was how to label wagons carrying dangerous goods. The chemists were concerned that different companies had different labels, and to avoid confusion it would be much better to have a common system of labelling. Jenkins of the Great Eastern was asked to design three labels, corresponding to the three additional categories in the special classification. He submitted prototypes, each bearing the name of the mythical "Great Scotch Railway" (see Figure 9.2). The labels were approved by the chemists and by the dangerous goods committee. Labels based on these prototypes were subsequently adopted and this would appear to be the first successful system in Britain for the uniform labelling of vehicles containing hazardous materials.

Of course, accidents involving dangerous materials did occasionally occur in spite of the regulations contained in the special classification. In such cases, the chemists proposed new or amended regulations to prevent a recurrence. In 1896, there was a spate of accidents in which consignments of charcoal caught fire during transit. There had been no injuries, but serious damage had been sustained by the wagons and in one case a signal box was destroyed. The chemists found that the problem was caused by the manufacturers sending the product by rail too soon after its removal from the retorts, the hot charcoal igniting in the brisk draught of air passing through the moving wagon. The chemists laid down minimum times for various types of charcoal to be exposed to the air before being transported.

The Railway Chemists in Collaboration 159

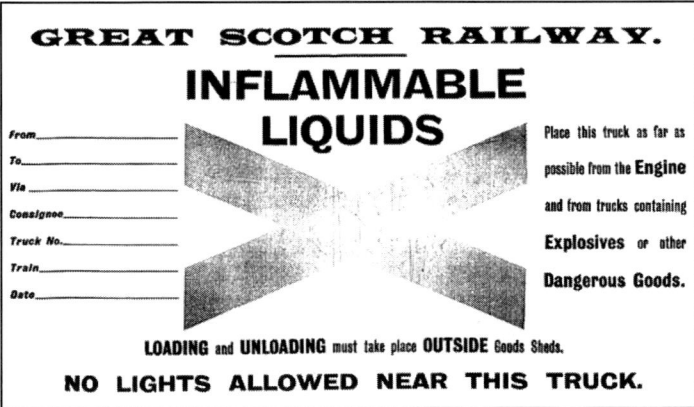

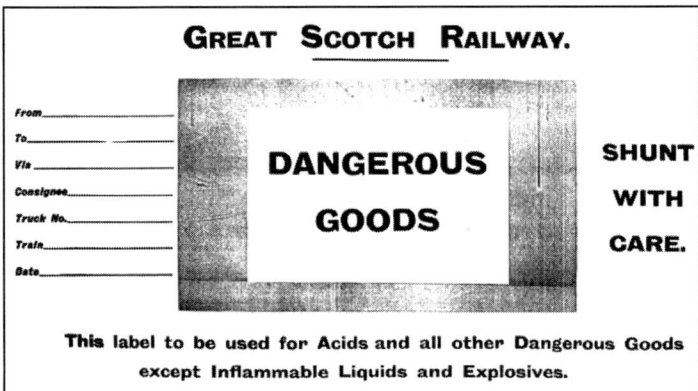

Figure 9.2 Prototype labels for hazardous materials prepared by Jenkins (chief chemist of the Great Eastern Railway) in 1914. The labels measured 327 × 176 mm. The centre design was in red. The Great Scotch Railway was a mythical creation. (BRB (Residuary) Ltd.).

In 1919 and 1920, two identical incidents occurred in which leakages occurred from drums holding a concentrated solution of a weedkiller containing an arsenic compound. In both cases the poisonous liquid was absorbed by sacks of sugar. Fortunately no deaths occurred, but in the first instance five people became ill after purchasing sugar from a grocer in Itchington, and in the second 60 people were struck down in Haslemere. The chemists ruled that consignments of all such materials should bear a label saying "Poison – Must not be loaded with Foodstuffs", and that arsenical weedkiller solutions should contain a dye, so that if contamination were to occur in future it would be easily detectable.

While most freight was carried by goods trains, packets and parcels were also transported in the guard's vans of passenger trains. The chemists advised on whether particular items could be conveyed by passenger train and, if so, what regulations were appropriate. For example, in 1902 it was discovered that a paper parcel of celluloid bicycle mudguards placed on the hot water pipe in the guard's van of a passenger train had started to smoulder. The chemists recommended that such pipes should be surrounded by metal gratings so that no article could be placed within a distance of four inches of them. Of course there was always the hazard (and indeed there still is) of the irresponsible passenger who takes a dangerous article on to a train with him. In his 1896 book on the handling of dangerous goods, H. J. Phillips recounted the story of an Irishman known as "Dynamite Dick" who earned a living by travelling the country breaking up blockages in blast furnaces by means of charges of dynamite. When Phillips asked him how he conveyed his dynamite, he replied that he often carried it with him in a portmanteau which he placed on the hat rack of the railway carriage!

While the bulk of the work of the committee concerned issues connected with the carriage of goods, the chemists were also called upon to advise on various other matters of general concern to the companies. One such issue arose in 1914 concerning indigo, which was used to dye the uniforms of railwaymen, and also those of members of the Army, Navy, Police and Post Office. The indigo plantations in India had suffered severely as a consequence of the introduction of synthetic indigo, and as a result a trade that had been worth between £3m and £4m in 1899 was worth only between £60 000 and £70 000 in 1914. Representatives of the vegetable indigo lobby were trying to reverse the situation, claiming that that natural indigo was a faster dye and that it was antiseptic, unlike the synthetic dye which contained powerful and dangerous acids and was poisonous. They also argued that natural indigo feeds the cloth, while synthetic indigo weakens it. The railway chemists advised that there was no difference between natural and synthetic forms of the dye itself, but that the natural material contained a small amount of some additional compounds as impurities. They likewise dismissed the other arguments of the growers, and pointed out that the difference between the two products was so slight that it was doubtful if a clause in the regulations insisting that natural indigo be used could be enforced. The synthetic dye came from Germany, and the chemists said that while that situation was regrettable, the solution was for this country to establish its own manufacture.

9.5 THE RAILWAY CHEMISTS IN WORLD WAR I

In World War I, the government exerted a controlling influence over the railways through the railway executive committee, which consisted of the general managers of 11 companies under the (nominal) chairmanship of the president of the Board of Trade, but in practice under (Sir) Herbert Walker of the London and South Western Railway. In 1915, the committee decided that "a Joint Standing Committee of Goods Managers, Superintendents and Chemists be appointed to consider all matters relating to the conveyance of explosives for H. M. Government". Just as previously, the goods managers' dangerous goods committee had delegated discussion of the more technical issues to the chemists, so the chemists involved in this new committee were soon meeting alone (the so-called "special chemists committee").

The frequency of these meetings increased as the war progressed. Not surprisingly, the chief work of the committee was to respond to requests from bodies such as the Admiralty that regulations for the carriage of dangerous materials should be relaxed and that some dangerous materials should be permitted on passenger trains. In many instances these demands were agreed to, but sometimes it was felt that the potential hazards were too great, even in a time of national emergency. Thus the committee refused to allow very volatile inflammable liquids to be carried by passenger train.

Among the new materials being carried (by goods train) were chemical warfare agents, and in 1917 the chemists were asked to consider whether small samples of such materials could be carried by passenger train. This request arose because of the need to send samples from the factory to the laboratory where they were to be analysed. The committee agreed with this request provided that the consignment was accompanied by a chemist or by a person qualified to deal with the liquids in case of accident. How anyone was expected to deal with a chemical warfare agent being released into the wreckage of a railway accident was not specified![7]

When the war was over, the work of rescinding the regulations which had been introduced during the emergency was commenced. In January 1919, the joint standing committee and the special chemists' committee were disbanded, and thereafter the business of returning the railways to peacetime practices for handling dangerous materials was transacted by the original dangerous goods and chemists' committees.

9.6 CONCLUSION

The collaborative work of the railway chemists at the Clearing House commenced in 1892, and grew to be a sizable commitment by the time of Grouping. Its principal impact was in the field of the safe transport and handling of dangerous goods. Over the 30 year period of 1892–1922, the chemists' committee discussed hundreds of issues, the vast majority of which had a bearing on the safety of the travelling public and railway employees. There is no doubt that the good safety record of the railways in moving, handling and storing materials that were of an explosive, flammable, toxic or otherwise dangerous nature owed a great deal to the work of the railway chemists meeting at the Clearing House.

REFERENCES

1. O. S. Nock, *Railway Race to the North*, Ian Allan, London, 1976.
2. P. S. Bagwell, *The Railway Clearing House in the British Economy 1842–1922*, Allen and Unwin, London, 1968.
3. Railway Clearing House, Classification of Goods by Merchandise Train, 1847–1879 (TNA, RAIL 1091/30).

4. Railway Clearing House, Goods Managers Minutes, 1891–4 (TNA, RAIL 1080/195).
5. H. J. Phillips, *The Handling of Dangerous Goods*, Crosby Lockwood, London, 1896.
6. Starting with this meeting (on 14.06.1893), the Minutes of the Explosives and Chemists' Committees were issued separately from those of the Goods Managers: Railway Clearing House, Chemists' Committee and Explosives and Dangerous Goods Committee, 1893–1923 (seven volumes) (TNA, RAIL 1080/808-14).
7. Railway Clearing House, Minutes of the Standing Committee of Superintendents, Goods Managers and Chemists appointed by the Railway Executive Committee to consider all matters relating to the conveyance of Explosives and other Dangerous Goods for H. M. Government, 1915–1919 (TNA, RAIL 1080/636).

CHAPTER 10

The Enduring Legacy of the Railway Chemists

10.1 GROUPING AND AFTER

The year 1923 is famous in railway annals. Following wartime restrictions, it put to an end the unbridled competition between all the individual railways, and became the year in which nearly all agreed to work together in one of four Groups. The unifying effects of the Railway Clearing House were still evident, but they paled into insignificance beside the effects of 1923.

The Groups established were (in order of decreasing size):

The London and North Eastern Railway (LNER),
The London, Midland and Scottish Railway (LMSR, or just LMS),
The Great Western Railway (GWR or simply GW), and
The Southern Railway (SR).

They lasted until nationalisation in 1948.

A very small mileage of standard gauge track was not included in any of the Big Four (as they were called), and most narrow gauge railways as well. Much has been written about these events[1], and

Early Railway Chemistry and its Legacy
By Colin A. Russell and John A. Hudson
© Colin A. Russell and John A. Hudson 2012
Published by the Royal Society of Chemistry, www.rsc.org

one consequence was that, from the end of 1922, the individual laboratories became regional laboratories of the new larger companies.

By the time of Grouping, 14 major railway companies had appointed railway chemists (although not always with that precise title) and established laboratories for the pursuit of chemistry in the railway interest. There were still over 100 other companies who had not done so, but most were quite small affairs by comparison, and some had been able to get their chemical work done elsewhere.

Of the Big Four, each had a railway chemist who had charge of all the laboratories in his Group, and none was more successful than the LMS. The LMS now had four laboratories: at Crewe, Derby, Glasgow and Horwich. At Derby, the time had come for the Midland Railway chief chemist, Archbutt, to retire, for he was then 63. His successor, T. H. Adams, was eventually appointed chief chemical analyst of the LMS. Although Lewis-Dale was by far the strongest candidate for the top job in LMS chemistry, having acquired a London PhD in 1924, he remained in charge at Crewe. At first no title of "chief chemist" was conferred, as it was generally understood that the chief mechanical engineer, Sir Henry Fowler, himself carried that title.[2] However, in 1933 Lewis-Dale *was* given the title of chief chemist, but he was then based in London with no proper laboratory.

Conditions of work had varied enormously in the pre-Grouping railway laboratories; thus some had over a dozen on the staff at any one time, while others managed with two or three staff, the smallest being Inverurie which had only part-time staff without proper qualifications. Elsewhere the qualifications of the staff varied, though chemical expertise naturally increased in later applicants. Many had studied at mechanics institutes or the like, and more than a few ended up with the coveted Fellowship of the Institute of Chemistry. After Grouping some transferred for promotion from one railway laboratory to another, an example being Dr G. E. Wilson from Glasgow to Crewe in 1946. At a lower level, some small amount of interchange continued most of the time. There was a certain new degree of specialisation in many laboratories, some persons being charged with water analysis (for instance) and others concentrating on metallurgy. At least one man had been known as "manganese", and this practice of each chemist concentrating on (and hopefully perfecting) techniques in one small area seems an eminently sensible use of resources. In the end each laboratory had some kind of territorial responsibility, though the emphases of neighbouring laboratories may have been slightly different.

Thus the chemical work continued, only now in four companies, often with more than one laboratory in each. The chief chemist's department of the LNER started with six laboratories: in Stratford (east London), Doncaster, Gorton, Darlington, Cowlairs and Inverurie. By 1940, however, they had just three: at Darlington, Doncaster and Stratford East. The Stratford laboratory ceased to exist in 1941 after it was badly damaged by enemy action. The number of samples examined each year ranged from 5869 in 1927, to 16 720 in 1936, and fell off slightly after that as the clouds of war started to gather. The first chief chemist was J. H. B. Jenkins, who was succeeded in 1930 by T. H. Turner as chief chemist and metallurgist. The department was from then on under the control of the chief mechanical engineer, Nigel Gresley, for whom about a third of its work was done.[3]

Progress within the four Groups was not the same, the most innovative being the LMS department which adopted a centralised managerial structure and had research centrally directed. It happened thus. Shortly after its creation the LMS, like its rivals, suffered from internal competition, in this case between the former London and North Western Railway (LNWR), Midland and Lancashire and Yorkshire Railways (L&YR). To allay this problem the railway enlisted the help of Sir Josiah Stamp who, though no railwayman, was to lead the company, particularly in budgetary and policy matters. One member of a committee that he soon established was the distinguished scientist Sir Harold Hartley. Under his guidance the laboratories were now given a research mandate. This was a question of business efficiency rather than scientific enquiry for its own sake.[4]

In 1933, the LMS set up the first railway research department, under an engineer, T. M. Herbert, and in 1935 new laboratories for chemical and other research were established at Derby. In spite of the new emphasis on research in the LMS, on nationalisation in 1948, the "main preoccupation" of the laboratories distributed around the newly nationalised railway system was still, as a later deputy director put it, "with quality control and trouble shooting... using, in the main, traditional chemical methods".[5] In 1951, however, the British Railways research department was

The Enduring Legacy of the Railway Chemists 167

established and the enlarged organisation contained divisions devoted to chemistry, engineering, metallurgy, physics, textiles and operational research, as well as a headquarters section based in London. Some of these new divisions were performing work in areas (*e.g.* metallurgy and textiles) that had formerly been the responsibility of the chemists.[6] The chemists continued to perform their traditional roles of quality control and troubleshooting, as well as participating in research projects. At that time there were still 11 chemical laboratories at various railway works around the country. In 1964, a large new Railway Technical Centre was opened at Derby. This provided facilities for engineering research, and enabled the Derby-based chemists to take over the whole of the building that had been provided for the original LMS research department in 1935. The subsequent reorganisation of 1971 created the BR research and development division, with a change to project-orientated divisions.[7]

These many alterations in administration were accompanied by changes in the subjects and direction of research. However, this much is true: the elements of continuity with the past were so obvious that railway research in the post-1923 era may be assuredly seen as a legacy of the early pioneers in railway chemistry. We conclude by noting a few of the most important trends in post-Grouping railway chemistry.

10.2 WORK ACCOMPLISHED

10.2.1 Water

We start here because this was one of the earliest and most recurring themes to be dealt with by the railway chemists in the age of steam. Beginning with confident analyses by young independent chemists acting as consultants, it continued with earnest correspondence on the relative merits of alternative natural sources (as in the correspondence of William Bouch around 1860).[8] Then came patents on water-softening by Archbutt and Deeley in the 1890s, followed in 1930 by the formation of a committee to oversee the precautions which should be taken in the day-to-day administration of railway water supplies, chaired by Dr P. Lewis-Dale, which led to the decision to install lime-soda plants capable of treating three billion gallons each year.[9] However, the awesome spectacle of a fast Anglo-Scottish express scooping up several thousand gallons of specially softened water in a few seconds from the water troughs situated between the rails every 30 miles or so, was to pass with the demise of steam.

The use of soft water in boilers was not confined to steam locomotives. For a short while, diesel-hauled trains had a small boiler for carriage heating and stationary steam boilers were common features of engine sheds for many years. Boilers continued in use in other industries, and by 1961, ICI (for example) evaporated about five billion gallons of water annually, and even produced a book on the basis of its internal documents, showing how indispensible chemistry had become.[10] A chapter on "Treatment of locomotive boilers" gives a succinct account of the position by 1960.[11]

There is, however, another aspect to water purification that, after World War II, became of major importance. By 1954, Fancutt, then assistant director (chemical services) BR research department, could write "From the outset, railways have found it necessary to provide water supplies of potable quality to a large number of premises throughout Great Britain", from remote signal-boxes and country stations, to whole communities (as at Crewe and Swindon).[12] At about the same time, another writer put it even more strikingly, "It is not generally known that the railways control more drinking water supplies than any other interest in the country".[13]

In all this provision, quality control and advice remained the responsibility of the chemical services. This had become so important that the British Transport Commission issued a printed *Memorandum on the Control and Safeguarding of Railway Drinking Water Supplies*, under the auspices of the research department, chemical services.[14] This takes extremely seriously all aspects of the subject, especially the protection of supplies. Its declared purpose was "to summarise the precautions which should be taken in the day-to-day administration of railway water supplies",[15] words clearly taken from the agenda of Lewis-Dale's committee.

By the end of the 19th century chemical examination of potential drinking water had become a fine art, both in and out of the railway service. Many undesirable components of drinking water (such as heavy metals like lead) were derived from industrial waste, and the chemists were quick to detect and eliminate these.

Gradually, however, it became apparent that it was not chemicals but bacteria that were the life-threatening constituents of some water, and could convey all manner of water-borne diseases (such as typhoid and gastroenteritis). They were, as one author put it, our secret enemies,[16] and they had to be detected and eliminated. Chemical techniques could be used, and the methods of Wanklyn and of Edward Frankland were employed in railway chemical laboratories, to show whether a sample had been contaminated by microorganisms. Usually such water was condemned as unfit for drinking purposes. The application, however, of chlorine disinfection was possible as a last resort.[17] Although this kind of water examination was strictly the province of the microbiologist, on the railways it was the chemical services who did the analyses and offered advice. That was the situation in the years of Grouping and beyond.

10.2.2 Lubricants

Lubrication of railway machinery (*e.g.* locomotives, coaches, wagons, points, *etc.*) had always been important for reasons of cost and safety. We have seen how animal fats were first used, and that gradually they were replaced by alkanes from petroleum, and that these posed quite difficult problems for chemists. By the time of Grouping these had been largely solved, and lubricant analysis had become a complex business. The whole question was fully discussed in the influential book *Lubrication and Lubricants* by Archbutt and Deeley, who were, of course, railway servants in the best sense of the word. The first edition gave a powerful resumé of the knowledge of the subject at the end of the 19th century.[18] The book was the contemporary standard work for anyone concerned with lubrication or dealing with the process generally. It provides an example of how a railway laboratory made a major contribution to the engineering industry in general.

By the time of Grouping, things had changed so much that yet another edition of *Lubrication and Lubricants* (the 5th) was necessary, and duly appeared in 1927. The authors condensed information on all aspects of lubrication and lubricants in over 600 pages, and included every conceivable method of testing. They provided a textbook for both engineers and chemists working for the railways, as well as other engineering concerns. As they say "the standard methods for the testing of lubricants published by the Institute of Petroleum Technologists and the British Engineering Standards Association have been included" – and simply needed following.[19] In the case of wagon grease, the chief chemists of numerous systems contributed information which showed considerable variation. Informants included H. Gripper (ex-Great Central Railway), F. P. Matthewman (ex-London, Brighton and South Coast Railway), W. R. Bird (GWR), J. H. Jenkins (LNER) and the authors (LMS).[20]

The years 1923 to 1948 in Britain were unquestionably dominated by steam traction (though much of the Southern Railway was electrified, there were still very few attempts to drive rail vehicles by internal combustion). As we have seen, the advent of a superheater had occurred at the very end of the 19th century, although in a country with cheap coal the savings involved were hardly worth the money, so very few superheated locomotives appeared for some years. They began to be generally accepted by about 1910.[21] When this happened, the higher temperature of moving parts required a different type of lubricant, and this was the subject of study by some railway chemists. As Churchward remarked at the time,

> It has been found on the Great Western Railway, in the course of a considerable amount of work with superheating, that the question of lubrication must be specially attended to, and unless the lubrication is not only effective and sufficient, but also continuous, there is certain to be trouble from high superheat.[22]

Suitable lubricants, capable of working in the cylinders at much higher temperatures than before were developed by the railway chemists. Success here, reinforced by similar developments in France and Germany, led to the general introduction of super-heaters in the most powerful locomotives of the 1930s, the Pacifics or 4-6-2s.[23] When one of those Pacifics, the LNER's legendary *Mallard*, established the (still unbroken) high speed world record for steam locomotives (126 mph in 1938), could its designer, Sir Nigel Gresley, have spared a thought for the chief chemist of his department and his lubrication specialist, J. I. Hill, who, with others, had made such a feat possible?

Analysis and fundamental research on lubricants continued until after nationalisation, but as steam locomotives were beginning to be replaced by diesels the problems naturally changed. Much activity took place at the British Railways research department. An early contribution dealt with the quality of crankcase lubrication oils in diesel locomotives,[24] while two slightly later papers focused on the spectrography of diesel sump oils,[25] the spectrographs being at Derby and Muswell Hill (London) laboratories. A review paper by S. Bairstow, then director of scientific services division, research department, British Railways Board,[26] considered the scope in the railway industry of the science of tribology, which is a general study of lubricants, and the moving systems in which they work.

By the 1970s routine steam haulage had been completely removed from BR standard gauge lines. The advent of large-scale diesel haulage introduced all manner of change in the railways. A review paper from Derby[27] disclosed the latest development in lubrication on BR, and, concluded that "BR will continue to be a major user of petroleum products in the foreseeable future" (including fuels as well as lubricants), and that these would need to

Figure 10.1 Apparatus for recycling lubricating oil. BRB (Residuary) Ltd.

be adapted for specific railway usage, claiming savings of £4 million. Chemical tests were even more necessary given a four-fold rise in the cost of petroleum products between 1974 and 1978. They included spectroscopy and viscometry, as well as an engineering technique called "shock pulse testing".

By the mid-1980s three Derby-based engineers described research on a fuel-efficient, multi-grade lubricant for diesel locomotives.[28] Though mainly involving trials with locomotives, the work also involved chemical tests in the laboratory, including a modified oxidation test. The importance of chemical monitoring (chiefly by spectroscopy) is stressed in another paper from Derby, confirming the proven value of such analyses over many years.[29] Finally, an undated memorandum from BR research (apparently from the mid-1980s) revealed that this chemical technique had been applied to High Speed Train locomotives, and had reduced repairs in one year valued at over £1.5 million.[30] In this field alone, chemistry had been abundantly justified in financial terms.

Now we turn to a diametrically opposite problem: adhesion.

10.2.3 Adhesion

To facilitate the motion between two surfaces that should be moving, we need lubrication in the short term or, in the long term, application of tribology. For railways, however, there is a situation in which we want two surfaces in contact not to slide over one another, but for the one to adhere to the other – to achieve the maximum, not the minimum, friction. This, of course, is what should happen between a rail and a powered wheel moving over it. Such are the properties of the two metals that usually a powered wheel grips, or adheres to, the surface beneath it. When it fails to do so, and there is a "skid", the wheel is in danger of developing a "flat", and of course there may be loss of control of the train. There are records of whole trains failing to stop in wet conditions and crashing through level-crossing gates like matchwood.[31]

On occasions it has been customary for the adhesion to be further increased by the addition of sand, and sanding gear has been a feature of steam locomotives since Victorian times. More recently, a liquid composition of about 50% sand ("Sandite") has been used by trains specially designed to clear the rail surface. The Sandite is said to look like brown toothpaste.

Figure 10.2 "Leaves on the line" R. K. Taylor, "Trees along the line", *Arboricult. J.*, 1982, **6**, 183, HEC Associates Ltd.

The real causes of poor adhesion are either leaking oil from the train, debris on the line (such as fallen leaves),[32] or a combination of both. The solutions are mainly matters of engineering design or environmental management, but the chemist has been useful in identifying the nature of the contaminants, measuring friction, *etc.*[33] Though hardly chemical, a paper on "Trees along the line", to the great credit of The Railway Technical Centre to which the author belongs, delineates an annual problem that regularly provokes the travelling public to either annoyance or hilarity whenever it occurs (see Figure 10.2[34]).

10.2.4 Materials Examination and Development

Brief mention should be made of chemical testing of materials by the railways after Grouping and before nationalisation (*i.e.*, between 1923 and 1948). Much routine testing of course continued, as the figures already quoted for the LNER laboratories suggest. Taking 1935 as perhaps the peak, we find the following are the numbers of samples analysed: oils and greases (3297), boiler control water (3102), iron and steel (1498), paint *etc.* (1294), copper and alloys (1050), and then many other samples in smaller amounts. During *that* year the LNER chemists made 885 visits and consultations (the only item which showed a continuous increase for the next few years, mostly to assess the validity of claims from customers that goods had been damaged in transit). The reasons for testing, as in former times, included: doubts about the quality of materials supplied; suspicions that some goods might have been damaged in transit by faulty packing; the need to evaluate products coming freshly on to the market; and the spoilage of goods in railway warehouses (*e.g.* the infestation of grain and other organic goods). Through the examination of copper, steel and other metals, the laboratories were able to analyse for trace elements, and sometimes to advise on new treatments, including the welding of steel rails.[35]

Similar work no doubt was done in the laboratories of the other three railways, thus in preparation for a new steel plant at Crewe in 1926, three analysts were appointed to deal with just steel. Paint research was continued, and a wax evolved which could be used in carriage cleaning plants that protected the paint in life and appearance.[36] However, as we have seen, the LMS was to take a new turning, and about 1933 it formed a research department, and initiated projects like microanalysis (especially of rubber products), corrosion (particularly in electric relays), timber and concrete. In those inter-war years it held six conferences in London.

After nationalisation the development of new materials continued. A paper of 1974 described research in the lightweight

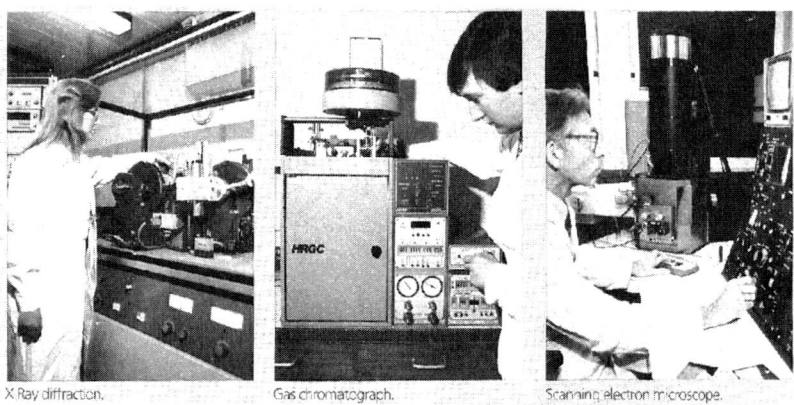

Figure 10.3 Analytical instrumentation in railway laboratories in the 1970s. BRB (Residuary) Ltd.

carbon-fibre-reinforced plastics (CFRP.), carbon-fibre-reinforced metals (CFRN.), and materials for a new sodium/sulfur battery, needed for electric trains when they had to move without drawing current from the overhead wires.[37] Other developments included use of structural adhesives in the construction of rolling-stock,[38] inhibited engine coolants[39], water-proofing for bridges and alternative compositions for paints and coatings.[40] Many other examples could be added.

10.2.5 Environmental Work

For many years trains spraying chemical weed killers have been employed to deal with the problem of unwanted vegetation growing on railway lines (see Figure 10.4). Although there is no evidence that this particular activity resulted in any significant environmental damage, as the importance of the environment became generally more appreciated in the post-nationalisation era, the railway laboratories rose to respond. Amongst the hazards sometimes presented by the railways to the public were those arising from their carriage of dangerous goods. An overview of the problem by the R&D division[41] presented the history of the problem and classified the hazards; outlined the legislation and specified correct labelling; detailed the design of tanker wagons; and much else. The chemists were heavily involved in monitoring what happened.

Obviously dangerous materials could be handled, but there were unknown hazards that travellers and the general public thought, rightly or wrongly, were associated with railway travel. As we have seen, the possible contamination of air in tunnels was early

Figure 10.4 A weedkilling train on the GWR in the 1930s. This sprayed a solution of sodium chlorate and calcium chloride on to the ballast.

pronounced absent, possibly because "these investigators were well paid not to frighten off potential passengers", as suggested in a much more recent paper.[42] This paper discusses the air quality in railway tunnels 135 years after the first such reports appeared (see Chapter 5). It lists the analyses of diesel exhaust (with carbon monoxide, sulfur dioxide and oxides of nitrogen the chief objectionable substances), and draws attention to the carcinogen, 3,4-benzpyrene, emitted by diesel and steam locomotives. It cites tests in the 1533 yard long Lees Moor Tunnel on the Bradford-Keighley line showing, amongst other things, the results when powerful steam and diesel locomotives with heavy loads passed through every 15 minutes (Table 10.1).

The author points out the significantly lower level of sulfur dioxide produced by the diesels, and goes on to claim that increased electrification helps to dispel the immediate problem for the railways, transferring it to the power stations.

With liquid pollution arising from spillage or carriage cleaning procedures, the monitoring of nearby water supplies became essential. From 1963 to 1971, engine fluids were treated with polychlorinated biphenyls (PCBs) to reduce their flammability, but the discovery of many thousands of dead seabirds was shown to be attributable to PCBs, which were thenceforth banned (though railways were far from being the only culprits). Solid pollution from a variety of sources, including waste deposited on railway land from neighbouring houses and industrial premises, was also found to be a problem but its disposal was not considered the responsibility of railway chemists. The use in railway carriages of asbestos called for constant monitoring by the chemists. The same was also true for toxic metals from many industrial processes carried out in railway workshops. The chemists continued to advise on safe working practices, aided by modern sophisticated techniques to deal with any problems that arose.

Table 10.1 Tunnel atmosphere – trains passing at high power every 15 minutes.

Type of locomotive	CO (ppm)	Oxides of S as SO_2 (ppm)	Oxides of N as NO_2 (ppm)	CO_2
Steam	up to 100	up to 15	Not detected	0.3
Diesel	up to 100	2	less than 5	0.1

10.3 CONCLUSION: THE END OF THE STORY?

We have seen how, since the 1830s, chemists have played a vital role in the development of the British railway system and in its safe and efficient operation. Initially they worked as private consultants, then as employees of railway companies. In this capacity they saw the major organisational changes of Grouping in 1923, and then nationalisation in 1948. The most recent change has been privatisation in 1996, when the four chemistry laboratories still operated at that time by British Rail (at Derby, Swindon, Doncaster and Glasgow) were sold to a firm of analytical consultants called Scientifics Ltd. The company provides a wide range of environmental and analytical services to industry in general, including the railway industry. Many other companies which supply the railway industry rely either directly or indirectly on the services of chemists and other scientists to develop and test materials used in the construction and operation of railway equipment. Examples include not only metals, lubricants, paints, and others which have made frequent appearances in previous chapters, but items such as the hydraulic fluid used in the tilting Pendolino trains and the fluorescent compounds used in the high-visibility jackets of track maintenance personnel. Chemistry is just as important as ever, but because chemists are no longer employed directly by the railway industry, their contribution is probably even less appreciated than formerly. The aim of this book is to put the record straight.

We conclude with these very few scraps of information on the fairly recent history of railway chemistry. Enough has been written to show that chemistry has produced immense changes. For example, setting aside all questions relating to the metallurgy and engineering of track, bridges and other structures, it must be obvious that locomotives would be very different without it. A steam engine that had to run on unsuitable water would remain like those in the age of Stephenson. No fast, long-distance trains would be likely and British social history would be utterly different. In the absence of appropriate lubrication, things would be even worse. The whole railway system (if one ever developed) would be unrecognisably different. Diesel traction would be impossible for similar reasons and also because the fuel oil would not be monitored. Electric traction might be feasible, though wagons and coaches would be primitive and inappropriate. Events like the railway races to the North would have seemed like science fiction, as would the concept of Pullman and dining cars.

Our final words are taken again from the article by G. E. Brown on "Railway chemistry", well over 100 years ago.[43] From the Swindon laboratory of the GWR he wrote:

> At first sight there seems but little connection between the silence of a chemical laboratory, with its array of shining glass vessels and delicate balances, and the roar of an express as it thunders through the night with its living freight. Yet... there is a very intimate connection, and the one has a good deal to do with the safe, efficient and economical working of the other.

Today one might notice the absence of instrumentation in such laboratories. Otherwise, these words are as true today as then. There will always remain a vital connection between the railways and chemistry laboratories. The laboratories, as the chemists who work in them, will surely continue to play a key role in the ongoing technological development of the railway systems of this country.

REFERENCES

1. See *e.g.*, D. H. Aldcroft, *British Railways in Transition*, Macmillan, London, 1968, ch. 2.
2. R. E. Jones, *Brief History of Crewe Laboratory*, privately circulated typescript, 1961 (TNA/DA, 69).
3. F. S. Lilly, *Notes on the Chief Chemist's Department, LNER*, privately circulated typescript, 1941 (NRM/DA, 423).
4. C. Divall, Down the American road? Industrial research on the London, Midland and Scottish Railway, 1923–1947, in *Railway Management*, ed. J. Armstrong, C. Bouneau and J. V. Olivares, Scolar, Aldershot, 1998.
5. K. G. A. Pankhurst, Research with British Rail, *Quest (J. City University, London)*, 1972, no. 21, 7.
6. S. Wise, *British Railway Research – the First Hundred Years*, ed. A. O. Gilchrist, Institute of Railway Studies, York, 2000.
7. B. J. Hawthorne and M. T. Hall, Platform for chemists: R&D on British Rail, *Chem. Br.*, 1978, **14**, 72.

8. *Railways: Scrapbooks of Early History, Letters and Cuttings, ca. 1813–1885*, vol. 1, ff 61 and 65 (NCL, L 656.2). See also SDLUR, Letters from Thomas Bouch, 1856–63 (TNA, RAIL 632/70 File 4).
9. Sir Harold Hartley, Scientific research on the London Midland and Scottish Railway, *J. Inst. Transport*, 1932, 495, pp. 502–3.
10. P. Hamer, J. Jackson and E. F. Thurston, *Industrial Water Treatment Practice*, Butterworth, London, 1961.
11. Ref. 10, pp. 291–306.
12. F. Fancutt, The chemist in railway service, *J. R. Inst. Chem.*, 1954, **78**, 517.
13. Anon, *Chemistry and British Main Line Railways*, July, 1947 (pamphlet apparently produced for a Science Museum exhibition) (NRM/DA 264).
14. Research Department, Chemical Services, *Memorandum on the Control and Safeguarding of Railway Drinking Water Supplies*, B.T.C., n.d.
15. Ref. 14, p. 1.
16. P. F. Frankland, *Our Secret Friends and Foes*, SPCK, London, 1892 (and subsequent editions); the author of this best-selling volume was son of Edward Frankland.
17. J. Race, *Chlorination of Water*, Wiley, New York, 1918.
18. L. Archbutt and R. M. Deeley, *Lubrication and Lubricants: A Treatise on the Theory and Practice of Lubrication, and on the Nature, Properties and Testing of Lubricants*, Griffin, London, 1900.
19. Ref. 18, 5th edn., 1927, p. vi.
20. Ref. 19, pp. 177–8.
21. W. A. Tuplin, *British Steam Since 1900*, Pan Books, London, 1971, p. 41.
22. Cited in H. A. V. Bulleid, *Master Builders of Steam*, Ian Allan, London, 1970, pp. 124–5.
23. E.g., C. J. Allen, *British Pacific Locomotives*, Ian Allen, London, 1962.
24. S. Bairstow, Control of quality of crankcase lubricating oils of locomotive Diesel engines in service, *J. Inst. Locomotive Engineers*, 1961, 98–112.
25. P. T. Corbyn, Experience on British Railways with spectrographic examination of used diesel sump oils, *Proc. Inst. Mech. Eng.*, 1963, 17–40; P. T. Corbyn and A. F. Haines,

Spectroscopic examination of diesel sump oils, *Railway Gazette*, 1965 (June 18th), 478.
26. S. Bairstow, Some examples of the scope of tribology in the railway industry, paper presented at the Industrial Lubrication and Tribology Exhibition, London, November, 1969.
27. A. F. Haines and G. R. Morley, Lubricant developments on British Rail, *Proc. Inst. Mech. Eng.*, 1978, **192**, 333.
28. G. R. Morley, J. A. Ellend and K. Dunn, British Rail switch to multigrade, *Ind. Lubricat. Tribology*, 1984 (July/August), 124.
29. G. R. Morley, Condition monitoring techniques on British Rail Diesel engines, *Maintenance Management Int.*, 1985, **5**, 127.
30. British Rail Research, *Analysis of used engine oil*, Research & Development Division, Derby, n.d. (mid-1980s).
31. For example, Bob Jackson, *Steaming Ambitions*, Triangle Publishing, Leigh, 1999, p. 44.
32. T. M. Beagley, I. J. McEwan and C. Pritchard, Wheel/rail adhesion – the influence of railhead debris, *Wear*, 1975, **33**, 141.
33. M. Broster, C. Pritchard and D. A. Smith, Wheel/rail adhesion: its relation to rail contamination on British Railways, *Wear*, 1974, **29**, 309.
34. R. K. Taylor, Trees along the line, *Agric. J.*, 1982, **6**, 183.
35. Ref. 3.
36. Ref. 9.
37. B. A. W. Redfern and G. W. J. Waldron, Materials research in British Railways, *Metallurgist Mater. Technologist*, 1974, (July), 305–310.
38. A. Seeds, The future use of structural adhesives in the construction of British Rail rolling-stock, *Int. J. Adhesion Adhesives*, 1984, **4**, 17–21.
39. A. Gaukroger, *Inhibited Engine Coolants: User Experience on British Rail*, R&D Division, Feb., 1980.
40. Ref. 5.
41. R. J. Jopson and A. G. Bale, *Dangerous Goods on British Rail*, R&D Division, October, 1984.
42. E. D. Henley, Some railway environmental issues, paper released from BR R&D Division, November, 1974 (NRM/DA, 605).
43. G. E. Brown, Railway chemistry, *Railway Magazine*, 1898, **2**, 58.

Subject Index

3,4-benzpyrene 176

Abinger Hammer, Surrey 25
Adams, T. H. 165
"adepts" term 37
adhesion 172–3
adulteration of oils 111–13
Ahrons, E. L. 132
air
    quality in tunnels 65–8, 104–6
    use by steam engine 30
"alchemists" term 37
Alkali Act 37
Allen, A. H. 81
Allen, C. J. 124
Allhusen Works, Gateshead 4–5
ammonium chloride 55
An Illustrated History of British Railways' Workshops 88–9
Aquadag (lubricant) 140–1
Archbutt, Leonard
    analysis 35–7
    animal/vegetable oils 142
    appointment 82
    calorimetry 104
    canister design 154
    claims for damage 119
    copper testing 102, 131
    curcas oil 147
    free acids in oil determination 111
    fuel analysis 104
    iodine absorption for turpentine analysis 115–16
    lead sheathing 137–8
    lubrication 18, 86, 97, 111–13
    Lubrication and Lubricants 18, 82, 169
    metallography 136–7
    oxygen in copper determination 102
    railway chemist 139, 156
    retirement 165
    rosin grease 140
    water softening 82, 143–5, 167
    work of railway chemists 95–6, 117–18, 121
atomic theory 31–3
Avogadro, Amedeo 33
Avogadro's Hypothesis 33

Bacon, Joseph 81, 86
Bairstow, S. 171
Bell, Isaac Lowthian 80–1, 135
bench reagents 98
benzene ring structure 34, 36

Berzelius, Jöns Jacob 35, 37, 38
Bessemer converter 85
Bessemer, Henry 73–5
Bessemer steel 74–5, 76–7, 80, 85, 99–100
Bethell, John 11, 64
Bird, W. R. 86, 156, 170
Birkinshaw, John 16
bismuth 131–2
bismuthate process 17, 101
"boiler compounds" 55
boilers
  corrosion 55, 106–8
  efficiency 7
  explosions 19, 45–6, 106–7, 133
  scale 44–6
  soft water 168
  tubes 45
"bomb calorimeter" 104
borax bead test 36
Bouch, William 52, 167
Boys of All Ages 5
Brass 17
Brighton and South Coast Railway 86, 170
Bristol and Exeter Railway 64
British Association for the Advancement of Science 3
British Engineering Standards Association 170
British Non-Ferrous Metals Research Association 138
British Railways
  lubrication 171
  petroleum products 171
  research 88–9, 166–7, 172
broken rails 100–1, 122
Bromsgrove 19
bronze 17

Browell, E. J. J. 65
Brown, G. E. 88, 178
Brunel, Isambard Kingdom 10–11, 14, 54, 62–64
Burghardt, Charles Anthony 68
Burt, Boulton and Harwood 65
Bury, Edward 49–50

Caerphilly Castle (locomotive) 21–2
calcium carbonate 44
calcium nitrate 157
calcium sulfate 44
Caledonian Railway 84, 86–7
calorimetry 104
Calvert, F. C. 53
Cannizzaro, Stanislao 4, 33
carbon chemistry 34–5
carbon dioxide emissions 21
carbon-fibre-reinforced metals (CFRM) 175
carbon-fibre-reinforced plastics (CFRP) 175
cast iron 16, 26–7
Chapman, W. G. 5
charcoal
  block test 36
  coke 15
  iron ore reduction 26
  transportation 158
chemical consultants
  air quality in tunnels 65–8
  conclusions 69
  fuel analysis 59–61
  preservation of timber 61–5
  swansong 68–9
  water testing 43–56, 59

Subject Index

chemical revolution
  atoms and molecules 29–37
  coal, iron and the rest 24–9
  scientific social
    revolution 37–41
Chemical Society 3, 39–40, 53
chemistry as "useful subject" 7
Chester College 75
chlorine 25
Churchward G. J. 106, 129, 170
claims for damage 117–21
Clark, Thomas 54, 143–4
Clow, Archie 24
Clow, Nan 24
coal
  calorific value 104
  coke conversion 28
  gas 15, 28, 104
  railways 20–1, 25
  sulphur 103–4
  tar 28–9, 62–3, 65
  testing 103
Coalbrookdale, Shropshire 26–7
Cochrane, Archibald (9th Earl of Dundonald) 28–9, 63
coke
  atmosphere
    contamination 21
  charcoal replacement 15
  conversion from coal 28, 60
  Darby I, Abraham 26
  fuel analysis 61
  fuel for *Rocket* 20
  Great Western Railway 60, 66
  smoke 20
"common carriers" 152

copper
  metallography 136
  Rocket 17
  testing 101–2
  tubing 129–131
copper sulfate 64
corrosion of electric relays 174
cottonseed oil 113
Cowlairs 166
creosote 62–5
Crewe Mechanics' Institute 76, 78, 94
Crewe and railway
  chemists 75–9, 104
Crookes, William 75
Culross, Scotland 28
curcas oil 147
Cutoxine (dog washing preparation) 119

Dalton, John 31–3
dangerous goods
  committee 155
  labelling of wagons 158
Darby I, Abraham 15, 26–8
Darby III, Abraham 27
Darlington 81, 166
Davy, Humphry 14, 38
Davy, John 66
Day, James 81, 86
Dean, William 17, 82, 131
Deeley, Richard M.
  lubricants/lubrication 18, 138–43
  Lubrication and Lubricants 82, 139–40
  water softening 143–5, 167
Department of Scientific and Industrial Research (DSIR) 141

Derby 8, 81, 143, 165–7, 171–2, 179
Die Modernen Therien der Chemie 4
diesel
　haulage 171
　　sulphur dioxide emissions 176
　　traction 177
dining cars 178
Dods, John 75–6, 86
Doncaster 105–6, 166, 177
DSIR 141
Duke of Wellington 45
dynamite
　carriage 160
　discovery 6
　Mount Cénis tunnel 6–7
"Dynamite Dick" 160

eggs (damage in transit) 120–1
elaïdin test (adulteration of oils) 113
electric railways 52, 137, 170
electricity supply 130
electroplating 116
"elements" 29–30
emergency cord 94–5
Engineering Chemistry 97–8
environmental work 175–6
ether (carriage) 155
Euston 78
explosives
　carriage 153–4
　committee 153, 155

Fancutt, F. 88–9, 168
Faraday, Michael 11, 62–4
Fellowship of the Institute of Chemistry (FIC) 76, 81

fireboxes 130–3
Flying Scotsman 110
Forth Bridge painting 114
Fowler, Henry 129, 165
Frankland, Edward 2–3, 24, 40, 53, 68, 108, 169
French Revolution 31
fuel
　chemical consultants 59–61
　coke and LNWR 61
　Rocket 20–2
　testing 103–4
Fuller, C. J. P. 86, 116
Furness Railway 68

gases 32–3
Gateshead 4–5, 81
Glasgow 84, 165, 178
glycerides for lubrication 110
"GNR standard" 96
Gooch, Daniel 10–11, 14, 60
Gooch, John Viret 50, 55
goods tendered for carriage 117–21
Gorton 166
Graham, Thomas 43, 53
Grantham 105
grease
　lubrication 113–14
　rosin 141
　specification 97
　tallow oil 114
　wagon 170
Great Central Railway 85, 144, 157
Great Eastern Railway 85–6, 156
Great North Scotland Railway 156

Subject Index 185

Great Northern Railway
  analysis
    cottonseed oil 113
    goods in transit 118
    materials 116
    miscellaneous 122
  chemists' role 105
  competition 151
  laboratory 85–7
  meetings 156
  water supplies 106
Great Scotch Railway
  (labels) 158–9
Great Western Railway (GWR)
  Boys of All Ages 5
  Caerphilly Castle 21
  coke 60, 66
  construction 62
  Dean, William 17, 82, 131
  Grouping 164
  Hackworth 14
  Kyan process 62
  laboratory
    analyses 116
    location 86–7
    meetings 156
  railway chemists 82
  research 129
  timber preservation 64
  water
    analysis 49
    softening 54
    supplies 106, 144
  weedkilling 175
Gresley, Nigel 166, 170
Gripper, H. 86, 112, 141, 157, 170
Grouping
  broken rails 100
  competition 89–90

  creation, 1923 164
  laboratories 85–7
  legacy of railway
    chemists 164–7
  organisational changes 177
  research 129
gunpowder 6–7

Hackworth (locomotive) 14
Hadfield, R. F. 99
Hall, Mike 8
"hammer-ponds of Sussex" 25
Harris, F. W. 77, 86, 97–8, 131
Hartley, Harold 129, 166
Henry, W. 46–7
Herapath, John 65, 68
Herapath, William 49, 61, 65
Herbert, T. M. 166
High Speed Train
  locomotives 172
Hill, J. I. 170
Horwich 165
Hübl, A. 112
Huish, Mark 74–5
Hull and Selby Railway 49, 63

ICI and water usage 168
iglodine 116
indigo 160–1
Industrial Revolution 24, 26, 45
inert gases 34
inorganic chemistry 34
Institute of Chemistry 40, 53, 64, 76, 79–80, 89
Institute of Civil Engineers 50
Institute of Mechanical Engineers 151

Institute of Petroleum
  Technologists 170
Inverurie 166
iodine value (oils) 112–13
iron
  production 15
  pyrites 103
  railways 25
  Rocket 15–17, 25
  see also cast iron; pig iron;
    wrought iron
Ironbridge 27

Japanese wood oil 146–7
Jenkins, J. H. B.
  Great Eastern
    Laboratory 85–7, 166
  Japanese wood oil 147
  labelling of wagons for
    dangerous goods 158–9
  luminous paint 147
  metallography 136
  pianos in transit
    damage 119–20
  railway chemist 156, 166
  wagon grease 170
  weathering of paint 146
"jerkers" 19
"jerkwater lines" 19
Joy, W. and E. 18

Kekulé, August 4
kerosene 105
"King Coal" 21
Kirtley, Matthew 60
Koch, Robert 109
Koettstorfer, J. 111
Kyan, John 62–4
Kyanizing (timber
  preservation) 62–4, 116

labelling of wagons for
  dangerous goods 158
Lancashire and Yorkshire
  Railway (L & YR) 68, 85,
  86, 106, 116–17, 154, 166
Lancaster and Carlisle
  Railway 51, 53
Lancaster and Preston
  Railway 2
Langmuir, Irvine 142–3
lard oil 110, 113
Lartington 51–2
Lavoisier, Antoine 31
lead sheathing 137
"Leaves on the line" 173
Leeds and Selby Railway
  65, 67
Leeds and Selby Tunnel 65–7
Lees Moor Tunnel 176
Lewis, E. A. 132
Lewis-Dale, Percy 79, 86,
  147–8, 165, 167–8
Liebig, Justus von 35–7,
  59–60
lighting 105, 147
"lime soda" and water for
  boilers 20
linseed oil 114
Literary and Philosophical
  Society of Newcastle 14, 80
Liverpool and Manchester
  Railway
  demand 15
  gradients 65–6
  opening 43–4
  rails support 62
  Rocket 12
  water quality 46–7
London and Birmingham
  Railway 49–50, 51, 63, 67

Subject Index

London, Brighton and South Coast Railway 86–7, 170
London, Midland and Scottish Railway
  chemical analysis 88
  competition 166
  Grouping 164
  research department 166–7, 174
  Stanier, William A. 82
London and North Eastern Railway
  Chemists' Council minutes 8
  Grouping 164
  laboratories 166, 173–4
London and North Western Railway
  Bessemer steel 74
  carriage photographs 78
  claims for damage 119–21
  competition 151, 166
  Crewe Mechanics' Institute 76
  fuel analysis 61
  laboratory 86–7
  lead sheathing 137
  Railway chemists 73, 82
  steel manufacture 99, 103
  steel testing 73–5
  Tipler, F. 97
  water
    analysis 73
    locomotives 19
London and Southampton Railway 63
London and South Western Railway 161
Losh, James 14
Lowestoft 85

lubricants
  testing 109–14
  work accomplished by railway chemists 169–72
lubricants/lubrication research 138–43
Lubrication and Lubricants 18, 82, 139–40, 169–70
lubrication and *Rocket* 18
Lucas, John 12
luminous paint 147

MacPherson Engine Blue paint 146
Mallard (locomotive) 170
manganese alloy steel 99–100
manganese in steel determination 76, 79, 101
Marecco, A. F. 64, 68
Markham, Charles 60
martensite steel 136
materials examination and development 173–5
Matthewman, F. P. 86, 170
Maumené value 112
Memorandum on the Control and Safeguarding of Railway Drinking Water Supplies 168
Mendeleléef, Dmitri Ivanovich 34
metallography 134, 136–7
metals, research 129–38
methane structure 34
"Metropolitan Mixture" 104
Metropolitan Railway 68
Meyer, Lothar 4, 34
microanalysis of rubber products 174

Midland and Great Northern Joint Railway (MGNJR) 68, 69
Midland Railway (MR)
  appointment of chemists 82
  broken rails 100
  competition 151
  copper testing 102
  fireboxes 20
  laboratory 86
  lead sheathing 137–8
  meetings 156
  research 129
  water softening 144
  white metal 102–3
mineral oils 110–12
Moon, Richard 74
Mount Cénis tunnel 6–7
Muswell Hill 71

National Railway Museum, York 8, 12
nationalisation of railways, 1948 177
Newcastle and Carlisle Railway 50, 51
Newcastle Central Station 99–100
Newcastle societies 14, 80
Nobel, Alfred 6
non-ferrous metals and *Rocket* 17–18
North British Railway 85–7
North Eastern Railway (NER) 64, 68, 102
  Bell, Isaac Lowthian 80
  laboratory 81, 86–7
  steel analysis 100
Northumbrian (locomotive) 45

oil gas 147
Oildag (lubricant) 140–1
oils
  lighting 105
  lubrication 110–14
  physical properties 110
olive oil 110–11, 113
open-hearth process (rails production) 85

paint
  luminous 147
  pigments 114–15
  research 174
  testing 114–16
  zinc white 146
paraffin oil 157
Pattinson, J. 37
Pease, Edward 48
Pease, Joseph 48
Pendolino trains 177
Peppercorn, A.H. 6
Periodic Table 34
"permanent way" 73–5
Peto, Samuel Morton 50
petroleum spirit 157–8
Phillips, H. J.
  carriage of explosives 153–4
  copper plates 131
  Engineering Chemistry 97–8
  Great Eastern laboratory 85–6
  railway chemist 77–8, 86
  railway chemists committee 154
  sulfur in copper determination 102
  The Handling of Dangerous Goods 154, 160

Subject Index 189

turpentine analysis 115
water testing for domestic use 108–9
Phillips, Richard 67
"phlogiston" 31
phosgene 66
photography 78
photomicrography 135
physical chemistry 35
pianos (damage in transit) 119–20
pig iron 15, 27, 73, 85
α-pinene 115
piston rods 15
Playfair, Lyon 2
poisons 160
polychlorinated Biphenyls (PCBs) 176
Porter–Clark process 144
preservation of timber 61–5
Priestley, Joseph 30–1
Primrose Hill tunnel 67
privatisation of railways, 1996 177
producer gas 104
"puddling" and iron production 15
Puffing Billy (locomotive) 11–12
Pullman cars 178
Punch 95

quality control (materials testing) 117, 122–3

Races to the North 89, 151, 178
rails
    production 8, 85
    quality 122
    testing 100

Railway and Canal Traffic Act 153
railway chemists
    Carriage of Dangerous Goods Act 152–4
    collaboration or competition 151–2
    committee 154–6, 157–61
    conclusions 162
    Railway Clearing House 121, 152–4
    World War I 161–2
Crewe 73, 75–9
materials testing
    analytical methods 97–9
    claims for damage 117–21
    conclusions 122–3
    copper 101–2
    fuels 103–4
    goods tendered for carriage 117–21
    introduction 94–6
    lubricants 109–14
    miscellaneous 116–17
    oils for lighting 105
    other activities 121–2
    other metals 102–3
    paints 114–16
    quality control 122–3
    specifications revision 122
    specifications and standards 96–7
    steel 73–5, 99–101
    water 105–8, 108–9
public image 88–90
work accomplished
    adhesion 172–3
    conclusions 177–8
    environmental work 175–6

railway chemists (*continued*)
    lubricants 169–72
    material examination and development 173–5
    water 167–9
Railway Clauses Act, 1845 152
Railway Clearing House
  carriage of goods 117–18, 121
  dangerous goods 152–6
  railway Chemists' Committee 155–6
  unification 164
Railway Magazine 6, 88
Railway Technical Centre 167, 173
railways
  competition 89, 151
    gauge 21
    laboratories, chemists/location 86–7
Railways Act, 1921 14
Railways and the Victorian Imagination 1
"railways without chemists" 1, 38
Rainhill Trials, 1829 12–13, 20–1, 38, 43, 95
Ramage, Hugh 76, 101
Ramsbottom, Jas. 19, 75–6
rare gases 34
Reddrop, Joseph 76–7, 86, 98–9, 101, 104, 109
Reid, David Boswell 67
Reindeer (locomotive) 55
Rennie, John 50
research on railways
  background 128–9
  conclusions 148
  introduction 128
  lubricants/lubrication 138–43
  metals 129–38
  minor projects 145–8
  water treatment 143–5
Reynolds, O. 149
Richardson, Thomas 37, 51–2, 54, 60, 61, 64–5
Ritterbandt, Louis Antoine 55
Roberts-Austen, William 131–2, 135–6
Rocket (locomotive) 10–23
  boiler scale 45
  chemical challenges 8
  copper 17
  diagram 13
  fuel 20–2
  iron 15–17, 25
  lubrication 18, 19
  mechanism 30
  modifications 12–13
  non-ferrous metals 17–18
  Rainhill Trials 43
  water 18–20, 19–21
  wrought iron 16
Rothman, R.W. 66
Routledge, George 81
Routledge, Robert 81, 86
Rowley, E.W. 81, 86, 116, 118–19
Royal College of Chemistry 39
Royal Institution 38
Royal Society of Chemistry 40–1, 53, 151
Royal Society, London 39–40, 48
rubber springs 117

salt 25
Salt Tax 38

Subject Index

Sandite 172
saponification value (oils) 111–12
Scheele, Carl William 31
Scheele's Green (copper arsenite) 116
Science Museum, London 12–13, 21
"Scientific Revolution"/ "scientific revolution" 2
Scientifics Ltd 177
Septonal antiseptic dressings 116
Settle–Carlisle line 151
signal lamps 68, 78, 145
Simmons, Jack 95
sleepers 29, 64
Smith, Angus 53
smoke pollution 20–1, 60
Society of Public Analysts... 37, 95–6, 117–18, 121
soda 20, 25
sodium peroxide 154
sodium/sulfur battery 175
softening of water 54, 82, 106, 143–5, 167
South Durham and Lancashire Union Railway 51–2
South Western Railway 55
Southern Railway (SR)
  electrification 170
  Grouping 164
spectroscopy 171–2
Spencer, George 117
Spinney, Thomas 61
springs 16
Stamp, Josiah 166
Stanier, W. H. 82
Stanier, William A. 82

steam engines
  air 30
  interest 5
  Rocket 16
  thermodynamics 2
  Tornado 6
  water, for 30
steel
  Bessemer 74–5, 76–8, 80, 85, 99–100
  chemical analysis 16–17, 99–101
  manganese alloy 99–100
  manufacture 16–17, 28
  martensite 136
  production 79
  specification 96
  testing 73–5, 99–101, 174
Stephenson, George 12, 16, 17, 21, 46, 177
Stephenson, Robert 12, 14, 16, 17, 50, 177
Stewart, James 76
Stockton and Darlington Railway 43, 48–9, 50
Stratford East (London) laboratory 85, 97, 115, 166
sulfuric acid 25
superheating 109, 170
sulfur dioxide emissions 176
"swallow's nests" (deposits in locomotives) 103
Swann, E. 75–6, 86
Swindon laboratory 82–4, 109, 178
sugar 119, 121

tallow oil 110, 113–14
The Age of the Railway 1
The Analyst 136

The Handling of Dangerous Goods 154, 160
The Railway Magazine 6, 66, 88
thermit 118
thermodynamics and steam engine 2
Thorpe, Edward 135
timber, preservation 61–5
Tipler, Frederick C.
  appointment 78–9, 86
  eggs (damage in transit) 121
  gas analysis 104
  grease 97
  laboratory accommodation 145–6
  metals analyses 133–4
  oil lamps 145–6
  railway chemist 156
  Swindon laboratory 82–3
Tornado (locomotive) 6
tramways 16, 27
"Trees on the line" 173
Trevithick, R. 16
Turner, T. H. 166
tunnels 65–8
turpentine analysis 115–16

United States (US)
  railways 2–3
  valence 33–4

valency concept 33–5

W. H. Smith 74
wagon grease 170
Walker, Herbert 161
wallpaper (arsenic content) 116
Wanklyn, James Alfred 108, 169
Warington, Robert 38
water
  hardness 19, 122, 143
  railway chemists 167–9
  research 143–5
  Rocket 18–20
  softening 54, 82, 106, 143–5, 167
  steam engines 30
  testing
    analysis 7, 46–53, 59, 73
    boiler scale 44–6
    conclusions 56
    dawn of railway age 43
    domestic consumption 108–9
    for locomotives 105–8
    treatment 54–6
    troughs 19, 165
Webb, F. W. 133–4
weed killers
  carriage 160
  on tracks 175
West, William 45, 47–51, 54, 60
white metal 17, 102–3
Wijs, J. J. A. 112
Williamson, James 66
Wilson, G. E. 165
Witham, Thomas 52
Workington 85
World War I and railway chemists 161–2
wrought iron 15–16, 26, 28, 99

York 81
Young, J. W. 105
Young, W. G.
  damage to goods in
    transit 118–19
  GNR laboratory 86
  miscellaneous analyses
    116–17
  oils for lighting 105
  specifications and
    standards 96
  tallow oil 114
  thermit 118
  zinc white paint 146

zinc chloride 64